A LIFE CROSSING BORDERS

Memoir of a Mexican-American Confederate
Las memorias de un mexicoamericano en la Confederación

Recovering the U.S. Hispanic Literary Heritage

Board of Editorial Advisors

A LIFE CROSSING BORDERS

Memoir of a Mexican-American Confederate
Las memorias de un mexicoamericano en la Confederación

REV. SANTIAGO TAFOLLA

EDITED BY CARMEN TAFOLLA AND LAURA TAFOLLA

ENGLISH TRANSLATION BY FIDEL L. TAFOLLA

Recovering the U.S. Hispanic Literary Heritage

Arte Público Press
Houston, Texas

A Life Crossing Borders : Memoir of a Mexican-American Confederate / Las memorias de un mexicoamericano en la Confederación is made possible through grants from the City of Houston through the Houston Arts Alliance and by the Exemplar Program, a program of Americans for the Arts in collaboration with the LarsonAllen Public Services Group, funded by the Ford Foundation.

Recovering the past, creating the future

Arte Público Press
University of Houston
452 Cullen Performance Hall
Houston, Texas 77204-2004

Cover design by Pilar Espino

Tafolla, Santiago, 1837-1911.
A Life Crossing Borders : Memoir of a Mexican-American Confederate / by Santiago Tafolla ; edited by Carmen Tafolla and Laura Tafolla; English translation by Fidel Tafolla.
p. cm.
"Published as part of the Recovering the U.S. Hispanic Literary Heritage Project series"—Intro.
Translation of original handwritten, Spanish-language manuscript entitled Memorias de un mexicoamericano en la Confederación; includes Spanish transcription and English translation.
Includes bibliographical references.
ISBN 978-1-55885-597-7 (alk. paper)
1. Tafolla, Santiago, 1837-1911. 2. Mexican Americans—Biography. 3. Mexican Americans—Ethnic identity. 4. Confederate States of America. Army—Biography. 5. United States—History—Civil War, 1861-1865—Personal narratives, Mexican American. 6. New Mexico—History—1848—Biography. 7. Santa Fe (N.M.)—Biography. 8. Circuit riders—United States—Biography. 9. Ranchers—Southwest, New—Biography. 10. Justices of the peace—Texas—Biography. I. Tafolla, Carmen, 1951- II. Tafolla, Laura. III. Tafolla, Fidel L. IV. Title.
E184.M5T35 2009
978.9'04092—dc22
[B]

2009038008
CIP

∞ The paper used in this publication meets the requirements of the American National Standard for Information Sciences—Permanence of Paper for Printed Library Materials, ANSI Z39.48-1984.

10 11 12 13 14 15 16 10 9 8 7 6 5 4 3 2 1

To Santiago

who dared to go searching for something better,

and to our parents,

who taught us to respect our heritage

as well as our dreams.

—C.T. & L.T.

TABLE OF CONTENTS

ix Introduction

1 A Life Crossing Borders: Memoir of a Mexican-American Confederate

91 Epilogue

107 Notes

125 Las memorias de un mexicoamericano en la confederación

206 Notas

208 Works Cited

211 Appendices

237 Acknowledgements

241 About the Translator

243 About the Editors

Santiago with 7 of his sons. Seated left to right are: J.J. Mercado, Jr., Santiago Tafolla, and Pete Tafolla. Standing are Santiago's youngest sons: Ernesto, Santiago Jr., Gabriel, Mariano F., Lott.

INTRODUCTION

Crossing borders is a daunting task. It requires the changing of rules, the changing of focus, and often, the changing of attitudes. It may also require the revamping of previously held beliefs. Santiago Tafolla, in the year 1908, at the age of seventy-one, took pencil to primary tablet and wrote about the borders, and the changes, of his life. This book carries the Spanish transcription, and the English translation, of that account and documents the life story of an individual who throughout his life crossed many borders—national, political, religious, cultural and personal.

Born in Santa Fe, New Mexico, Republic of Mexico, in 1837, to a father who had lived and served at the same presidio when it was under the flag of Spain, and then under the red, white and green of the newly born República de México, he would himself see and live through many changes of flag, traveling from the newly acquired-by-war U.S. Territory of New Mexico to the United States proper, from Missouri to New York, and from Pennsylvania Avenue in Washington D.C., to the plantations of Georgia. He would see Maryland, Kentucky, Ohio, Texas, and the Indian Territory of Oklahoma as part of the U.S. Army Second Cavalry, and subsequently as part of the Confederate Army. He would live in Mexico, in the state of Coahuila, as an expatriate for a year, then return to the United States as a rancher, a fur trader, a justice of the peace and finally, a Methodist preacher.

A personal witness to a time period of continually shifting borders, identities and allegiances, this is a powerful autobiography, challenging the stereotypes and reclaiming the history of Latinos in the United

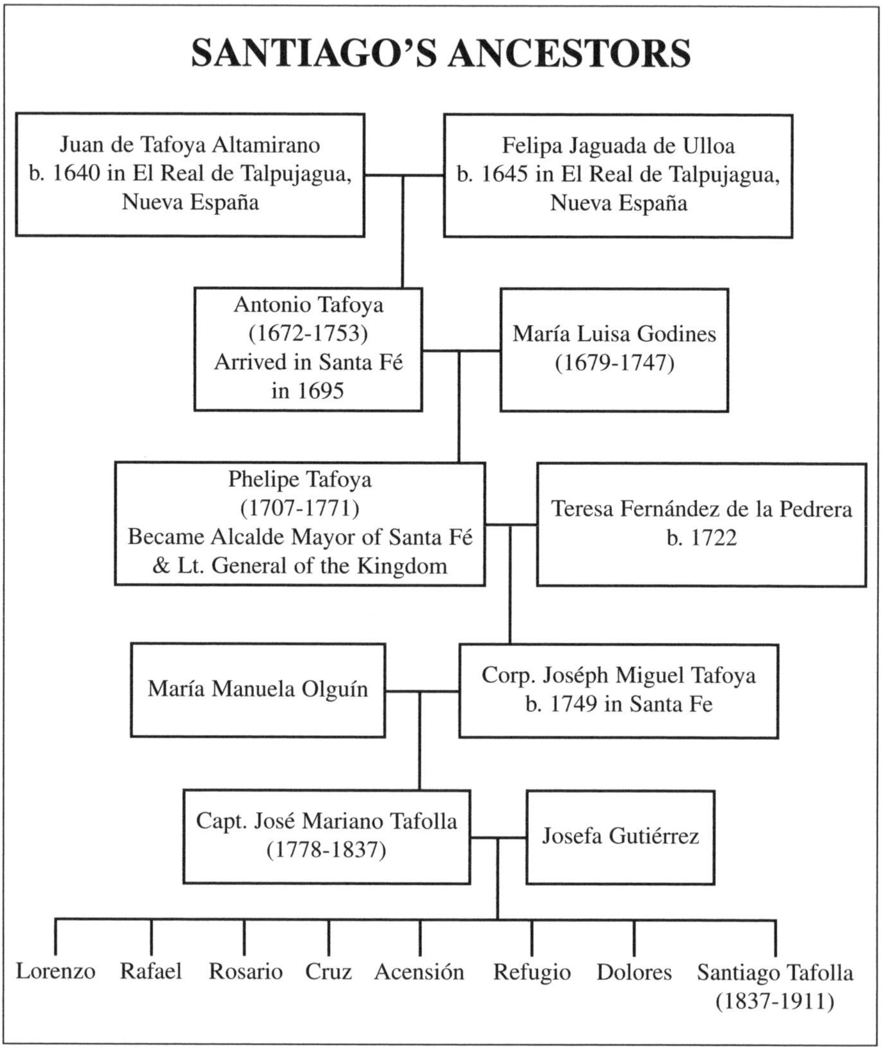

Ancestors of Santiago Tafolla, through the paternal lineage.

States. A valuable and fascinating work on its own, the historical links and substantiating documents that accompany it make it an even more important source for the understanding of our history and culture on this continent. Beyond the casual references to Robert E. Lee, Jefferson Davis, Gen. Juan Cortina, Congressman José Gallegos and Major Earl Van Dorn, the personal observation of lesser-known events, attitudes, biases and cultural conflicts adds to our understanding of such diverse situations as White Confederate gangs threatening to lynch their *mexicano* fellow Confederates; a wedding reception at Mormon Town, Texas; skirmishes between rowdy recruits from St. Louis and a Black crew on a Mississippi steamboat; and the Sunday afternoon goings-on at the residences of foreign ministers in Washington, D.C., where Santiago lived for two years, a few blocks from the White House.

As editors, we were surprised by Santiago's willingness to undertake the unknown, his eagerness for competence and learning in all areas and, having often been treated unfairly, his deep hunger to become a loyal part of an ethical institution. We were most struck by the courage, self-reliance and innovation shown by a boy who was the nineteenth century equivalent of a streetchild, as he struck out for new worlds, new nations, new careers. We believe that part of the reason he took his job as soldier so seriously, as well as a primary cause for his "pride" and indignation at mistreatment was this very experience with injustice and a threatened survival. Instead of conditioning him to injustice, it seems to have left him with an even deeper hunger for justice and even higher expectations of what life could and should offer him and his fellow human beings.

Given the many dangers and threats he faced, even Santiago himself seems surprised by his survival, and the word Providence appears many times in these pages, as he attributes this survival to the Providence of God, who "had some job cut out for me."

In a time of changing identities, when he is challenged as to whether he is White or Black, where he is surprised to be instructed not to associate with "Negroes," where he witnesses the existence of Indian-owned Black slaves who speak no English but only the language of their Indian "owners," Santiago Tafolla recalls finding himself—along with his fellow Mexican Confederate soldiers—terrorized by White Confederates. When the *americanos* take up arms and threaten to eliminate all the "greasers," Tafolla is forced to desert to Mexico in order to survive. He experiences firsthand the racism inherent to the time period, and even then attempts to build bridges between ethnic groups, often serving as the messenger or the communicator between ethnic groups, and praising the role his German immigrant friend Fred Metzger serves as peacemaker. A veteran of the Texas Indian Wars, though never given the lands he felt he had earned, he seeks out conversations with the Mexican captive of the Indians, and questions him about the internal discussions among the Indians. His story provides us priceless insights into the details and daily lives of people living "on the color line" and between the ethnic groups, and to the great extent to which family and personal relationships aided in survival.

The roles he serves vary from orphaned chore boy to runaway, U.S. wagon train waif to hotel messenger on Pennsylvania Ave., shoemaker apprentice, tailor apprentice, Army bugler, camel keeper, Confederate bugler, Confederate deserter, interpreter, rancher, farmer, Indian fighter, traveling salesman, trader in stolen hides, justice of the peace, civil rights organizer and for the last thirty-five years of his life, one of the first Methodist preachers of Hispanic descent in the United States, in a time period when the overwhelming majority of Hispanics were Catholics.

There are many different ways this book can be analyzed. As a young, orphaned runaway, later apprenticed to a variety of unethical or difficult "masters," he will surely be viewed by some as a real-life Mexican-American Lazarillo de Tormes or Oliver Twist, in a non-fiction version of the picaresque tradition. Dr. Norma Mouton, in her research and analysis of the autobiography of conversion in its Latino applications, views Santiago's memoir as classically fitting the first two of three requirements of the classic nineteenth-century novel of conversion. The major factor missing for its full qualification is the description of "asuntos de falda" or sexual adventures. Santiago, while admitting illegal ventures, pride, military desertion and a period of serious indulgence in "whiskey", becomes very shy when discussing his intimate relationships, nearly qualifying as Victorian in his exclusion of any romantic details, telling us simply that he promised marriage to a relative of Policarpio's, and later, in his second marriage, that he "renewed my acquaintance with Señora Anastacia, the widow of Corporal Mercado . . . " and the following month, married her. Fortunately, his second wife, Anastacia, the great-grandmother of the editors, lived well into her nineties and left the family quite a few detailed, romantic stories about his attentions, jealous actions and behavior during their courtship.

From an educational perspective, Santiago's story reveals a young man hungry for schooling, who makes several references to and criticisms of different educational settings, having attended lay schools in New Mexico, public and Catholic schools in Washington D.C., county schools, trade apprenticeships and a private theological institute in Georgia, and, we believe, seminary and theological studies in the Methodist-Episcopal church as an adult. (He is also, tellingly, referred to in his obituary as Dr. Tafolla.) The fact that we have available both the original account *and* an authentic translation by a grandson who knew him, had heard the stories orally in both languages and knew intimately the idiosyncracies of both his archaic New Mexico/Texas dialect of Spanish and his Southern dialect of English gives the reader options for better understanding Santiago's intention. This also allows a psycholinguistic analysis of bicultural experiences and expressions. For instance, in his uses of Spanish vs. English terms in different emotional settings—is there a specific time in his life when he uses *varas*, and other times *yardas* to describe distances? Are these tied to historical or national identity shifts? Is the term *varas* more frequently used when he is feeling besieged or threatened as he did while living under his brother's abuse in New Mexico? When and why does he switch to English prepositions or short phrases, as he occasionally seems to, almost unconsciously, in this Spanish-language memoir?

In addition, this memoir, written in retrospect by a Mexican-American Protestant minister living in a time period when to be Protestant was considered equivalent to being insane by most of the Mexican-American community, will surely be studied from the perspective of religious history, theological applications and ethical development. Studies of cultural identity, national identity and bilingual biculturalism will also find exciting new material here.

Santiago's role in the early awareness of ethnic prejudice and injustices, and his leadership in an 1870s civil rights meeting through a Masonic vehicle are also significant and revealing. This provides evidence of a meeting of Mexican-Americans who organized to fight social injustice, long before the commonly cited activism of the twentieth century.

It is important to note that this Santiago Tafolla, Sr. is not the same Santiago Tafolla, Sr. frequently referred to in the Handbook of Texas as one of the founders of the Order of Sons of America, School Improvement League, Cruz Azul Mexicano and other Mexican-American Defense Organizations of the 1920s and 1930s. That Santiago was Santiago Tafolla, Jr, born in 1872 and the son of our Santiago here. Santiago Jr. then fathered five children, among them a son also called Santiago (James, 1898-1966) who became a judge well-known throughout San Antonio in the 1950s as "Judge Jimmy." This latter James also had a son named James, who died during World War II at Iwo Jima.

The manuscript Santiago wrote is surprisingly honest. For a 71-year-old, well-respected Methodist Reverend with a reputation to uphold, he is quite open in his self-criticism, as when he relates that "liquor became such a vice to me that I was controlled by the liquor habit . . . I had gained the friendship of all the wicked men in the neighborhood and was well loved by all." He is also openly emotional, calling the night his mother-in-law died as "one of the saddest nights I've ever experienced in my life," empathizing with the young calf they had to kill for food, and still lamenting the needlessness of his first wife's death.

Very personal idiosyncracies become obvious in an account of this length. Two in particular appear repeatedly here: his love of horses and his hatred of intoxication. Santiago was an obvious lover of horses, sometimes to the neglect of commentary about the rider, e.g., "I had loaned my horse to one of my companions . . . When my horse got back . . . " and *"se me salía el corazón por lazarlos"* ("I wanted to rope them so badly that my heart was bursting out of my chest.") It is poignant and almost comical to see the great attention his memoir devotes to horses, and to his separation from, or reunion with this very valuable frontier companion, vehicle, and financial asset. In fact, it is quite telling from a religious point of view to know that after his conversion and calling to become a minister, he mortgages his ranch (and the horses there) in order to be able to afford the trip to move his family to a preaching circuit in Laredo, Texas.

The very negative and symbolic role which intoxication serves in Santiago's life is also obvious, first as a cause for fear when as a child his older brother would return drunk and angry, and later, as a major factor in judging others and himself for imbibing or not imbibing. Santiago, toward the end of his memoir, calls himself "malvado", wicked, for letting whiskey take over his life to the extent that he would disrespect laws and the original ownership of stolen goods, so much so that he runs for the office of Justice of the Peace, supported enthusiastically by those who wish to see him turn a blind eye to their illegal transactions.

It is at this moment that he pauses, at the lowest point in his life, when he, someone who has always prided himself on his own sense of right and wrong, makes a solemn oath he knows he does not intend to carry out. Perhaps he pauses in order to consider the way in which to best describe what comes next. Unfortunately for the reader, he does not finish, and we are left to wonder about the nature of that period of transition. We have accounts of the significance of his life after 1876, and he has told us his own story of his life previous to 1876. But that very moment of change and transformation, is what is left to the readers' and the scholars' imagination.

One of the characteristics of this document which most surprised us was that he appears to have written only one draft of the memoir. By all appearances, he sat down and wrote with little self-editing. A very few times, he seems to have read back a few pages, and added a letter or a chapter number, and a few points in the original manuscript show evidence of a new or sharpened pencil. The document is thus, possibly, written in several sittings. Most impressive is the level of educated biliteracy which he achieved, as a waif and an outsider, in a time period when few people were literate. His facility with the spoken and written language, honed, certainly, by his years as an evangelist, were also aided by the variety of educational systems in which he participated, from military to parochial, and public to theological, in both Spanish and English. He also shows evidence of self-education as a young boy, and mentions using Sundays in Washington, D.C. to read books and to visit the embassies of Mexico and of Spain. We have testimony that his sermons were moving and life-changing, and Methodist records state that even as a beginning preacher in a church where membership was 45, attendance was 250. His ease with storytelling was obvious, and this account seems amazingly spontaneous.

The journey which the publication of this book has taken, one hundred years in the making, deserves some telling here, though the epilogue and the acknowledgements carry much more detailed information about those who made this happen. The handwritten, Spanish-language manuscript—left unfinished when Santiago Tafolla died at the age of 74—was passed from relative to relative, spending much time in an antique vault, until his grandson, Fidel Tafolla, took on the task of translating and transcribing it in the late 1960s. Various copies of Fidel's transcription (smoothly edited into a more standard and modern Spanish) and its accompanying English translation circulated around the family and at the Nettie Lee Benson Archives until the original handwritten manuscript was found in an old shirt box in Fidel's closet, years after his death, and Laura (Fidel's daughter) and I (Fidel's niece) set about the process of verification, re-transcription and editing.

Our edited Spanish transcription attempts to stay true to the original manuscript. Except for standardization of spelling and punctuation, few changes were made, and for the benefit of history and authenticity, the flavor of the original vocabulary and phraseology was maintained. Hours were spent verifying and re-verifying (and magnifying) certain words, lines and phrases from a crumbling and faded pencil-on-tablet original.

Where questions occur, they are addressed in the Spanish notes. Historical links and background information, where relevant or helpful to understanding the context of the tale, are also added in the notes at the end of the English translation, but not included in the Spanish, so as not to distract from the flow of the original.

The editing of the English translation followed two loyalties: one, that it be loyal to Santiago's original document, emotional tone and intention, and two, that where discrepancies with the original text do not arise, that it be kept as close to Fidel's original translation as possible. Fidel's insight into this document and into Santiago's story has been invaluable. While Fidel's research was conducted from 1967 to 1971, and ours from 1986 to 2009, we defer to his appreciation of and personal contact with Santiago. As Mr. Wayne Vick so aptly stated, "Fidel's translation ITSELF is a historical document."

Published as part of the Recovering the U.S. Hispanic Literary Heritage Project series, *A Life Crossing Borders: Memoir of a Mexican-American Confederate / Las memorias de un mexicoamericano en la Confederación* is an invaluable aid to understanding the upheavals of the nineteenth century in North America. An absorbing account of personal survival in a world of fluid and changing borders, it is also an affirmation of ethnic identity in a time when racial and ethnic differences were subject to greater ignorance and often, greater violence. It is a story of hope and survival, of human struggle, indignance and pride, of reflection, compassion and indeed, as Santiago himself often affirms, of Providence.

—*Carmen Tafolla*

June 14, 2009

(One hundred and forty years to the day, after the marriage of Santiago and Anastacia, my great-grandparents.)

A LIFE CROSSING BORDERS

Memoir of a Mexican-American Confederate
Las memorias de un mexicoamericano en la Confederación

Santiago Tafolla, San Antonio, c.1908.

PEARSALL, TEXAS
SEPTEMBER 16, 1908

Inasmuch as my brethren in the faith have encouraged me to write my biography and as my children have asked me to do so, I hereby venture to undertake the task. I realize my inability to write properly for publication, because I know clearly that I lack sufficient education, but with the indulgence of my friends and desiring to grant their wish, I will propose to give them my biography, even though imperfect, to the best of my ability, hoping that those who read these poorly written lines will note that all I write is the truth, and I write this without any bias, not seeking honor, but, on the contrary, trying to portray all the circumstances which God in His Providence has allowed me to experience.

I begin my biography today at the age of 71 years and twenty-three days. I thank God that having passed through so many dangers, He has permitted me to bear witness to the many blessings which, through His Divine Grace, He has showered upon me.

When I first got to Texas, there wasn't another Tafolla in the whole state. Now I have seven sons, five daughters, and 45 grandchildren living here.

—Santiago Tafolla

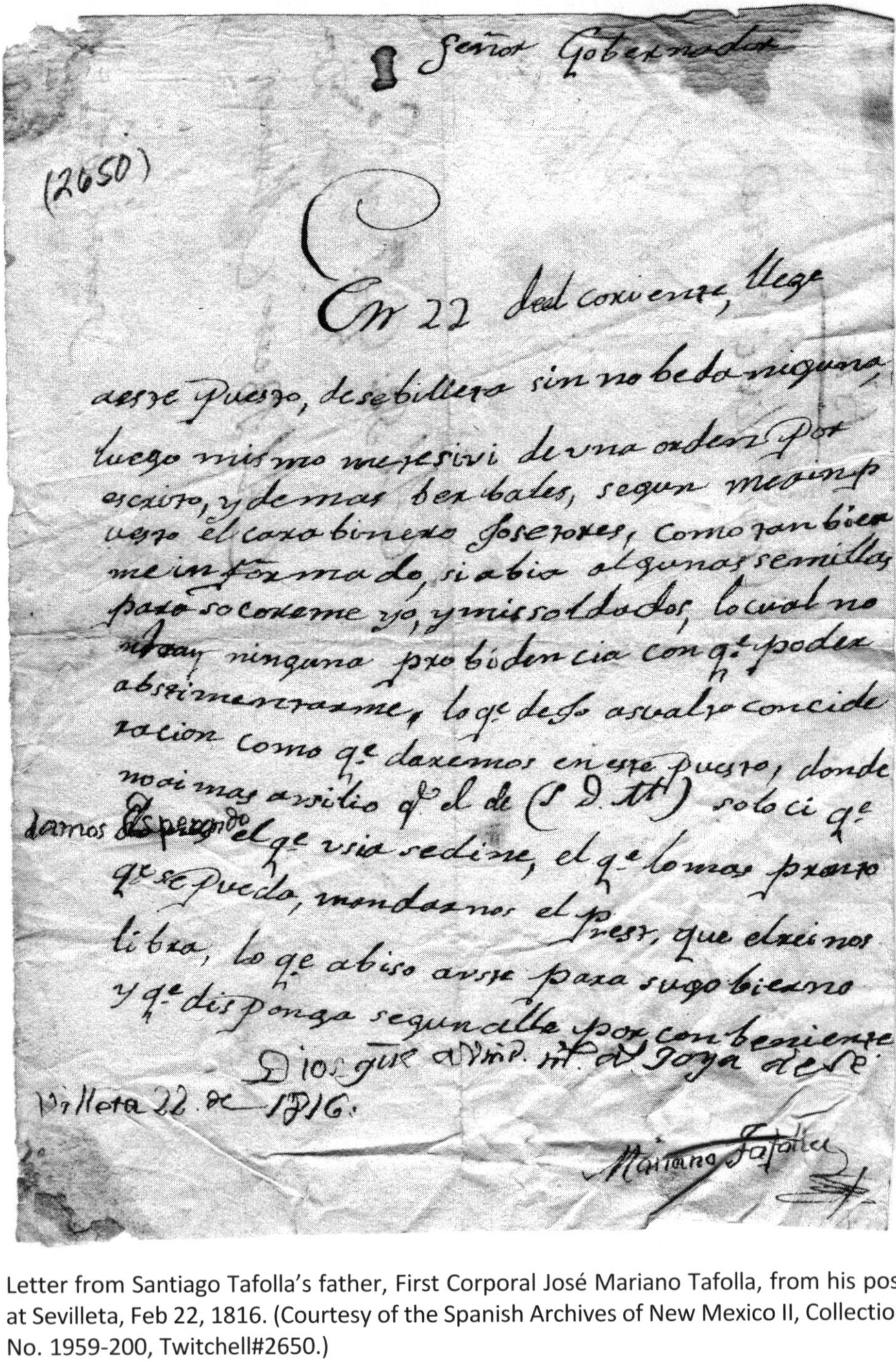

Señor Gobernador

(2650)

En 22 deel corriente, llegé
a este Puesto, de sebilleta sin no beda niguna,
luego mismo me resivi de vna orden por
escrito, y demas berbales, segun me ymp
uesto el carabinero Jose Torres, como tan bien
me informado, si abia algunas semillas
para socorerme yo, y mis soldados, lo cual no
[illegible] ninguna probidencia con qe poder
abstimentarme, lo qe deso a su alta concide
racion como qe daremos en este Puesto, donde
no ai mas ausilio qe el de (S D. M.) solo ci qe
damos esperando el qe vsia sedine, el qe lo mas pronto
qe se pueda, mandarnos el Prest, que el rei nos
libra, lo qe abiso a vse para su gobierno
y qe disponga segun allo por conbeniente
Dios gue a Vmd. m. a. [illegible]
Villeta 22. de 1816.

Mariano Tafolla

Letter from Santiago Tafolla's father, First Corporal José Mariano Tafolla, from his post at Sevilleta, Feb 22, 1816. (Courtesy of the Spanish Archives of New Mexico II, Collection No. 1959-200, Twitchell#2650.)

CHAPTER 1

I was born in Santa Fe, New Mexico, on July 23, 1837. My father's name was Mariano Tafolla and my mother's name was Josefa Gutiérrez, both natives of Santa Fe. My father was a military officer in the Mexican Army. He was returning from a journey to the capital of Mexico in a wagon train under his command that was carrying a large sum of money. While they slept at a place called Galisteo,[1] some thirty or forty miles from Santa Fe, the wagon train was attacked by Indians. The only one left dead at the scene of the encounter was my father. The group of forty or so men that made up the escort fled to Santa Fe to report the tragedy.

The troops sent to investigate the case found my father dead with an arrow piercing his heart. There were those who said that the very same members of the escort had killed my father in order to steal the money.[2] They were tried in court but I was too young to remember anything about it. All I know is what I heard discussed later by my mother and my siblings. I was only three years old when my father died.[3]

When I was seven my mother died and my oldest brother took in all the unmarried family members, which consisted of my adult sister and myself, the youngest member of the family. My oldest brother was named Lorenzo. I also had another brother, named Rafael, who was next in age to Lorenzo.[4] I had five sisters, four married and one single. Those married were named Rosario, Cruz, Acención and Refugio. The unmarried sister was named Dolores.

CHAPTER 2

In those days Santa Fe was known as La Villa and people in the surrounding villages referred to the people of Santa Fe as *los villeros*. The Tafolla family lived at a place called El Realito.[5] They owned property there, and their fields were irrigated. My brother planted various vegetables, but the main crops were corn and wheat. He always had three or four *peones* who worked the fields, and he made me work right along with them. He also had four or five burros. During the winter season he would send the *peones* to bring wood from the hillsides and he would make me go with them.

We would leave the house at about two or three o'clock in the morning on the coldest days imaginable. The ground was covered with snow, for in that part of New Mexico, snow covers the earth from November until early April. My work in connection with the *burricos* (that's what they called them) was to gather all the wood they cut, to guard the burros from wandering off, and to do whatever else they told me to do. And I had to do this in a hurry even though I was freezing. If they got angry with me, they would kick me and knock me down, and with each kick I would go sliding over the ice for about ten feet. All my bones would ache as if broken but there was no one I could appeal to for sympathy, for my brother was just as cruel to me as any of them.

When we returned to the village from the woods, about ten in the morning, the *peones* would go to their homes and I would go to sell the wood. As wood was scarce, it could be sold for as much as a *peso* a load; I had to sell it before I could go home. On some occasions, it was noon before I disposed of the wood and it was not until then that I had breakfast.

For the rest or the day, I was the one who was made to run errands for the home. If I delayed in returning from an errand, my older brother was so cruel that he would tie me, take off my shirt, and whip my back with a horsewhip until the blood flowed. The welts this caused would remain for a long time and were so painful that I could not lie on my back.

After witnessing one of these scenes, my sister who lived in the same house would meet me when we were alone, take me in her arms and cry with me. "Don't cry, dear sister," I would say to her, "Just wait till I get a little older and I'll run away." Hearing this made her cry all the more. I must have been about ten when she married and left the home. After that, I was left with no one to sympathize with me. My brother's wife didn't care much about my troubles. They had a daughter by the name of Francisca and since she was their only daughter, my brother and his wife adored her, and sometimes she was the cause of my suffering.[6]

CHAPTER 3

My brother was a respectable man and though not wealthy, he lived very comfortably. [Like my father], he had also been an officer in the Mexican army. He was what they call in Romanism "a religious man." He was very devoted to the Virgin of Guadalupe, and every year he would begin to pray the rosary for nine days prior to the twelfth of December, the anniversary of this virgin. And every night he would host the novena at his house, and invite many people. The house would be filled to capacity. After the rosary and the novena, he would offer a dinner to all the guests, and on the night of the twelfth, they would hold a vigil and pray till midnight. And then the event concluded with a dance where participants celebrated until dawn.

My brother had the habit of performing a charitable deed on the twelfth of December, and would choose two elderly people from among the poor of the community. He would load them down with corn and beans and portions of everything he had, and he would also give them pieces of muslin, fine fabric[7] and clothing.

In this he was very religious, but he was also a gambler who sometimes got so interested in this pastime that he failed to come home all night. When he won, he would come home very happy, but when he suffered a loss, he would go on a drunken rant. As soon as we heard his shouts, the whole family scurried about, looking for places to hide, as we knew he would be extremely angry. He would always commend himself to the Virgin of Guadalupe in order to win. Sometimes when he lost, he would get drunk with his own sadness. On such occasions, he would come home in the middle of the night, beating on doors, shouting and cursing. And as soon as we heard him, we began trembling with fear. For on these nights, he didn't remember his religion, and he didn't remember the Virgin of Guadalupe.

I was reared under such circumstances until I was eleven years of age.

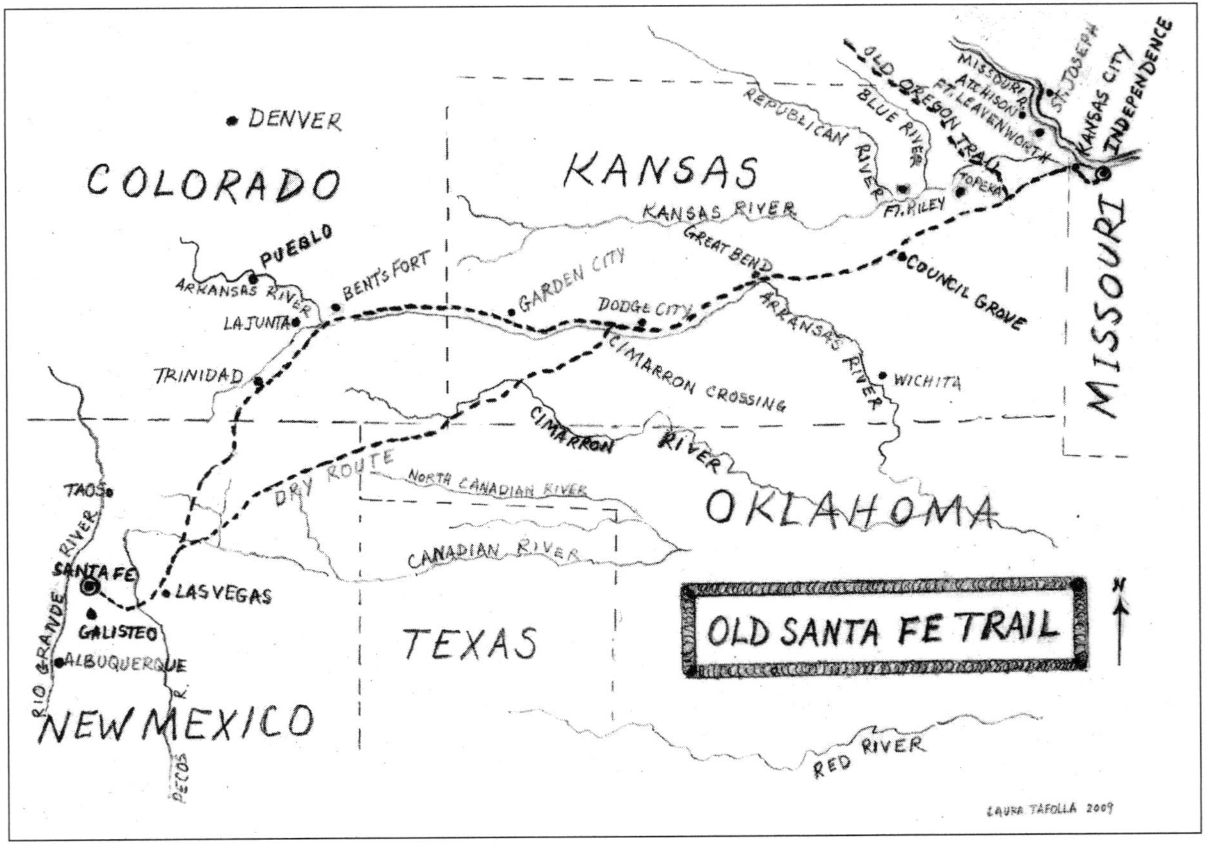

The Santa Fe Trail (approx. 780 mi.), showing the route which Santiago traveled from Santa Fe to Council Grove, Kansas, and Independence, Missouri when his wagon train came through in the fall of 1848. We do not know his exact route thereafter, but we do know that he went through St. Louis, Cincinnati, Ohio; Buffalo, New York; Baltimore, Maryland, and to Washington, D.C. (Map, Courtesy of L. Tafolla)

CHAPTER 4

In the month of September of 1848, my cousin Plácido and I got together and made some plans. His mother was my mother's sister, and he lived with relatives and led a life similar to mine. We decided to run away from home and escape to Las Vegas,[8] a town some seventy or eighty miles from Santa Fe.

The day we started, we walked about twenty miles and slept in the woods without any blankets or anything to eat.

We started walking and at about ten o'clock came to a spot along the road where a wagon train had camped the night before. We found some of the fires still burning. We picked up leftover pieces of bread and bacon, and that was our breakfast. Then we took to the road again.

After traveling five or six miles, we came to a place where the road had to make a wide circle skirting some hills. We figured that by cutting across we could catch up with the wagon train. We took a path which was used only by people on horseback or walking. It led across some very rugged terrain and several ravines.

Upon reaching the place where the path intersected the road, we were surprised to run into the wagon train passing right in front of us. It was an American wagon train which had departed from Santa Fe the day before. There must have been at least fifty wagons in the train and each wagon carried three or four men as passengers, some held more. This wagon train was owned by several men, but there was one individual who acted as commander. They had gone from Independence, Missouri, to Santa Fe to deliver government freight to the troops stationed in New

Mexico,[9] crossing the Staked Plains,[10] which at that time was a desolate area. The plains were covered with buffalo, and you could see thousands of them. There were also many deer, antelope and wild mustangs. And there were also many wild Indians, who were at war with the whole civilized world. This made it very dangerous to travel in those desolate areas at that time, so much so that the wagon trains would travel in two parallel paths, separated by a distance of about ten yards[11]—one-half the carts on one side of the trail and the rest on the other. In case of an attack from Indians, the front carts in both columns would come together and the carts in the rear would crowd in, forming a corral. The people would get on the inside of this enclosure which served as a fort. And they would form a corral in this fashion every night.

CHAPTER 5

When my cousin Plácido and I saw the wagon train, we stood paralyzed with fear. We didn't know what to do until a man in one of the carts said in very poor Spanish: "Machacho, ¿quiero vamos Estados Unidos?" I answered him, "Sí, yo quiero." I then nudged my cousin and said, "Vente, vamos."

But he stood there dumbfounded, for he was afraid of the *americanos.*[12] I went and jumped on the wagon. I kept calling my cousin, motioning for him with my hand, until I could see him no more. And that was the last time I ever saw my cousin.

The man who invited me to get on his wagon was named Robert A. Matthews, and from that day on, he took care of me. It took us more or less two months to travel across the Staked Plains.

During the journey, I witnessed many strange sights and experienced a number of adventures. On those prairies, I saw large areas populated by prairie dogs, and the places where they were most densely gathered were called "dog-towns."

We had in our train twelve Shawnee Indians. They were employed by the owners of the train as hunters and explorers and rode behind and in front of the train as scouts.[13] Each one of them was equipped with a rifle, the kind that back home were called *carabinas de pelo.* They were good marksmen. Every day they would bring in deer and antelope and also buffalo.

The prairie dogs would come out of their burrows, stand on their haunches and bark at the train from both sides of the road. One day, one of the Indians shot at one of them and wounded him so badly that his guts spilled out. Curiosity prompted me to bring the animal back to the place where we camped that night. And because I was foolish enough to be playing with him, he grabbed my finger with his teeth and would not let go until my companions took him off of me. I suffered for quite a while from that bite. My hand and my arm swelled up. Later, I learned that the bite of a prairie dog is quite poisonous, and that they share their burrows with rattlesnakes.

The train did not travel on Sundays. On one of those Sundays, a group of *americanos* went to bathe in a pond, and I went with them. But I didn't go into the water, because I was embarrassed to strip naked in front of them. The water was deep enough for swimming, and there were plenty of rush-like grasses, of the kind that grow in the water, so I didn't think it was that deep. When the men left, I threw myself in the water and ended up where it was very deep. Since I didn't know how to swim, I thought I was going to drown in that water. A thousand thoughts raced through my mind all at once. I thought I was going to die and that this was a punishment from God for having run away from my brother. Many other thoughts coursed through my mind, but I didn't lose consciousness. I grabbed the grass and pulled myself to the bottom of the pond. Dragging myself through the grass, I moved along underwater until I came to where I could stand. I lifted my head up, out of the water. I've never forgotten this experience.

After about a month and a half or two, we reached the first American homes on the Kansas border. From there, we took a stagecoach; I remember that the first American town I saw was called Council Bluff.[14] It was on the banks of a river. As I couldn't speak English, I didn't learn the name of the river, nor did I learn the names of the places we passed on the way. And I was also too young to be interested in such things. We arrived at about eleven in the morning.

While we were at the hotel waiting for the stagecoach, I witnessed another incident that I've never forgotten. The driver of the stagecoach stopped in front of the hotel. He got off and entered the waiting room. He took a money bag full of coins and began, or so it seemed to me, to insult all the men in the waiting room. There must have been at least fifteen or twenty people there. It appeared that he was trying to challenge everyone, for he walked up and down the waiting room with the money bag in his hand and shook it in the face of each person present, and no one dared to say a word.

There was a man sitting in a chair who apparently was sick, for he looked pale. The driver approached him and, pushing the money bag in the man's face, spoke in such a furious tone that the man became more and more pale by the minute. A tall, thin elderly man was sitting close by and he also looked sickly and pale. He had a thick, solid cane in his hand. On one of the exits the driver made from the waiting room, this old man walked to the door and waited for him. Just as the driver returned to the door, the old man dealt him such a blow between the eyes with the walking stick that the driver fell back on the rocky pavement. As the coachman tried to get up, the old man once again dealt him another blow on the head and made him fall behind his coach. The old man followed him and rained blows on his back. The driver would cry out at each blow but the people who gathered about would say, "Hit him, hit him again." Finally, the old man left him alone, not having uttered a single word from the time he dealt the first blow until he stopped hitting him. The driver then climbed to the driver's seat and took the reins in his hands. Two ladies who were already in the coach began to scream with fear. Several men ran to hold the horses by their bridles and others helped the ladies out of the coach. The coachman left whooping and hollering.

Fifteen minutes later, he returned with a companion and all the passengers, including me, boarded the coach. I didn't know what was going on except for what I could see, but what I heard, I didn't understand. That night, we arrived at Independence, Missouri, a very pretty town.

At the hotel, I encountered a Mexican who asked me how I liked the town. I replied that I liked it very much and he answered, "Don't become enchanted, young man, this is nothing compared to Mexico."

CHAPTER 6

The following day we left Independence. In the coach were: myself, Mr. Matthews, three other men and a young lady. At noon we transferred to another stagecoach at some town whose name I do not remember. That night we got to a very high and dangerous road and we were going up a very steep grade. The four horses that pulled the coach were very fat and spirited, and they began to balk before we had traveled halfway up the hill. It was decided that the men should get out and walk, and the young lady and I should remain in the coach. The men went on ahead, certain that the coach would catch up with them by the time they reached the top of the mountain. The young lady and I remained in the coach.

I held Mr. Matthews double-barreled gun in my hands. The horses pulled the coach a few paces, stopped and would go no further. They balked and began to rear up, and one of the two drivers, realizing that the coach was in danger of turning over, came and told us to get off and try to catch up with the men who had gone ahead. I had the gun in my hand, and the young lady and I went up the road hoping that we would catch up with them, or that the coach would catch up with us. It must have been twelve o'clock at night, and we had walked about a mile when we arrived at the summit of the hill.

Neither the young lady nor I had said a single word to each other. All of a sudden she stopped and said to me, "Will you take care of me?" Not understanding what she was saying to me, I answered, "No, sir." After a moment she walked over to a nearby oak tree, and leaned against it. I also stopped about ten yards from her, wondering what to do. I assumed that the young lady was frightened and that I had given the wrong answer to her question, but there was no way to explain myself, as she could not understand me any better than I could understand her.

We must have been in this perplexing situation for about fifteen minutes when one of the coach drivers rode up on horseback and, in some manner, made me understand that something had happened to the coach, and that I should go and break the news to the men ahead of us. I ran as fast as I could in the darkness of the night and, about a half a mile down the road, I found our companions standing, waiting for the coach. I made them understand as well as I could that something terrible had happened. The young lady's father was quite upset and began to ask me if anything had happened to the young lady. Not being able to give an explanation or to answer his questions, I looked at Mr. Matthews, who spoke a little Spanish, and he asked me in poor Spanish "¿Está mujer malo?" I answered "No, señor, nada le ha sucedido a la señorita."[15]

Then the father calmed down, and together we all returned to the coach. We discovered that it had turned over and rolled down the side of the mountain about a hundred yards and had been completely demolished. We stayed there until dawn, warming ourselves by the fire. The Americanos kept a steady conversation. I listened but was not able to understand anything. About eight o'clock the next morning another coach arrived to pick us up, and we continued our journey.

In those days, this particular region of Missouri was very thinly populated. Upon arriving at the next small town, we found the stagecoach with four horses waiting for us. There, two women got on. The young lady mentioned previously, and her father, got off there. During the day we arrived at a place where there was much sand, and the men got off and walked, as they were accustomed to doing when the road became difficult. I should mention that the two men were Mr. Matthews' companions and had been with him ever since we left New Mexico. One was named Love and the other McGoffin. They were our companions the entire trip, until we arrived in Washington.

CHAPTER 7

The ladies who were traveling in the stagecoach with us kept laughing and saying something about me. They kept looking at me, and I realized that they wanted to talk to me. And I kept thinking, "If they *do* speak to me, what am I going to say?" Just as I'd thought, one of them plunged in and asked me, "Do you speak English?" I answered, "No, sir." Upon getting such an abrupt and inappropriate answer, they burst out laughing, and I was somewhat embarrassed. Seeing my predicament, one of them answered, "That *is* English." As a matter of fact, that was all the English I knew.

To keep from making this part of my story too long, I'll just say we traveled by stagecoach for about ten days before arriving at St. Louis, Missouri. I don't know what date it was when we arrived. That night we slept in a hotel and stayed there for two days. The first morning after we'd arrived, about ten o'clock in the morning, I decided to go for a walk down the street. I had walked about three blocks when I heard the hoarse whistle of a steamboat that was about to reach the wharf. Never having heard such a sound before, I took off running down the street toward the noise.

When I got to the river,[16] I saw ships lined up on both sides of the river, and it seemed to me to be the most remarkable sight I had ever seen in the whole world. And truly it was, for I had never in my life seen so much as a rowboat. For more than two hours, I gazed in amazement at the loading and the traffic on the river. There were about fifty large steam ships, and many schooners and small sailboats. I was so intent in watching all this, that when I thought of returning to the hotel, I didn't know which direction to take, and since I knew neither the name of the hotel where I was staying nor how to speak English, I couldn't even ask anyone. I was lost until about three o'clock in the afternoon.

Finally, I arrived at a barbershop where I saw a Negro man who was a quadroon.[17] He looked like a Mexican, so I spoke to him in Spanish, but he didn't understand me. However, he took me by the hand to a house where there was a Mexican woman. He told her to ask me if I wanted to live with him, and that he would teach me the barber trade. I explained to the Mexican lady how it was that I had come to St. Louis. She asked the Negro to leave me there, and said she would see that I got to the place where I belonged. And then she advised me not to associate with Negroes and sent me to the hotel with an American boy. Mr. Matthews was very worried and had been looking for me.

From St. Louis, we took a steamboat to Cincinnati, Ohio, then to New York City and finally to Baltimore. I remember that we visited these places and also Buffalo, New York, but I never learned what business Mr. Matthews had at hand. As a matter of fact, the three partners I have mentioned arrived together at Washington, D.C. Mr. Love had a brother there who was a congressman from some state in the Union.[18]

We stopped at The United States Hotel. Mr. Matthews made arrangements with a Mr. Hackney, the owner, to give me room and board for whatever services I could render. I don't remember what month it was, but I think it was December when we arrived in Washington. I remained at the hotel for about ten months and during this time I learned enough English to perform my duties and do what I had to do. I became a favorite with the owner of the hotel and the rest of his family. I ran errands for all the guests and gained their friendship. My job was to be a messenger, to answer the calls when hotel guests rang for service. I knew the room numbers by heart and I made some money from the errands.

At that time, some rich gentlemen from Mexico City came to the hotel. I believe they were either on government business or were very rich because they spent plenty of money. They would hire me to interpret for them on certain matters, even though they had an interpreter. These gentlemen were Mr. Antonio Arrasol and his son, also named Antonio, and another gentleman whose surname was Balgán (I do not remember his given name.) The interpreter was named Juan Arrasol; he was an elderly gentleman who was very highly educated. These men set themselves apart from the other guests, for they would ask to have their table set separately and they would order choice foods and fine wines at all hours. They were very liberal with their money. Sometimes they would give me five dollars for running an errand, and if there was a dollar or two left in change, they would give it to me. I became a favorite with this group.

I did not have to bother Mr. Matthews for clothes, for I earned enough from my errands to buy my own clothes. About this time the office man who had the night shift at the hotel left his job, and I was given his position. This also paid me earnings from the guests arriving or leaving at night. During the time that I held this job, I didn't have to do the messenger job, so I had all day free to sleep. I took advantage of this opportunity to attend a public school during the day.

Around this time, the presidential elections took place and President Pierce[19] was elected. He ran against General Scott, the general who led the forces of the United States in the war with Mexico in 1846 and 1847.[20]

CHAPTER 8

I had been living at The United States Hotel in Washington for about ten or eleven months when Mr. Matthews and Mr. Hackney, the owner of the hotel, had a disagreement. Mr. Matthews moved out of the hotel and took up residence elsewhere. He was working at the Capitol, on certain congressional matters; I never learned just what his business was.[21] When he moved, he rented a room at a boarding house managed by a lady named Mrs. Mary A. Davis. She rented six or seven furnished rooms. The boarding house was in front of the National Hotel on Pennsylvania Avenue, on the same street where The United States Hotel was located.[22] Some members of Congress were staying at the boarding house.

Mr. Matthews moved me from where I'd been, brought me here and arranged for Mrs. Davis to employ me as an errand boy and whatever other duties I could do, on the condition that, during school hours, I would attend school. Mrs. Davis was a very good woman. Her elderly mother lived with her, as well as her two children, a boy of six and a girl of eight. The cheapest rooms rented for a dollar a day, but she got as much as $2.50 per day for the better rooms. Rooms in Washington were very expensive in those days.

Mrs. Davis recommended me to a Catholic school that was close to Georgetown. I do not know why she chose to do this. I do not think she was a member of any church, nor did I ever see her attend church or have any religious worship at home. Nevertheless, her little girl attended a

Catholic school and perhaps she thought, since I was Mexican and Catholic by birth, that this school was where I belonged.

The only teachers I knew were the teachers I had known in my homeland, and they were all laymen or *majoros*.[23] This was a school with 3 or 4 classrooms. They put me in the classroom for the youngest children. From the very first day I noticed that the style of teaching was different from the public school where I had been. Most of the time was spent praying. The first thing they would do in the morning was go to Mass, which was celebrated in a chapel on one side of the same building.

I have never seen men so cruel with children. In my classroom, there were some children who couldn't have been more than six years old. And they would punish these children so cruelly that it made me recall my own childhood at the hands of Lorenzo. For the least thing, they would punish six- or seven-year-old children. At times they would grab them by their ears and lift them three feet off the ground and throw them, and the children were not to cry. I was among the oldest children in that classroom, and I made up my mind that the day they punished me like that, I would leave school.

Mrs. Davis paid fifty cents a month for me to attend. I suffered a few small punishments that were not too severe because it seemed they were treating me with some consideration, and I thought maybe that was because I was Mexican. But one day, I don't know why, the layman[24] called me up from my seat. I had done something, I don't remember what, but he grabbed me by the ears (which seems to have been his favorite punishment) and he didn't lift me up off the ground because I was already too heavy. But he left my ears red.

That evening, I told Mrs. Davis that I no longer wanted to attend school and I told her what was happening there. The next morning she handed me fifty cents and she wrote a note to the layman telling him I no longer wished to go to school. All I had learned there in three months was how to pray the Salve, the Ave Marias, the Our Father and the Apostles' Creed, all in English. I learned nothing about reading.

After that, I returned to the public school where I had been previously. I stayed at Mrs. Davis' house about a year. During this time, I met the Plenipotentiary Minister of Mexico. On Sundays they would give me books after I finished my chores, and I would get a chance to visit the minister, who would receive me kindly. He had some young men there from Mexico City who were attending *colegio*,[25] and they would treat me very well and joke around with me. And they would call me el Nuevo Mexicano because I was from New Mexico.

I also visited the Minister of Spain. I went to his house, and he received me well. He asked me several questions, and then he asked me if I was visiting him to obtain something from him. I thanked him and bid him farewell. His name was Mr. Calderón. I think the Mexican minister's name was Matías Barrientos.

CHAPTER 9

Mr. Matthews married about this time and decided to take his wife on a visit to his birthplace at Talbotton, Georgia.[26] He took me as far as Charleston, South Carolina, and left me with his brother there who was a bricklayer. I was there about three months and then his brother sent me to another brother who lived in Muscogee County.[27] His name was Charles Matthews. He was a doctor who owned a plantation with several slaves. His wife had a school of the kind known as "County Schools." There I attended school for about three months.

After school and on days when we had no school, I worked at whatever chores they assigned to me. One day there was a fight at school, and the other boys accused me of causing the trouble. They told the doctor, and he came and scolded me in a way that was trying to scare me. I told him that he didn't scare me, for I was only scared of one. He then asked who that person might be, and I told him God. Upon hearing this, he smiled and left me all alone in the school building, which was about fifty yards from his home.

I felt offended and wondered what I should do. I was ashamed to go back to the home, so I made up my mind to leave the place. It was about eight o'clock at night. The railroad called the "Muscogee Railroad" passed near there, and I had heard others say that it was only twenty-five or thirty miles to Columbus.[28] So I followed the railroad tracks. I walked all night long and the next day too. I didn't have anything to eat except some peaches from the orchards which grew alongside the tracks.

That night I arrived at Columbus about eight o'clock. I followed a street and after a while came to a hotel and sat down on some benches in front of the building. Since there were lots of people there passing, entering and leaving the hotel, no one took particular interest in me. I fell asleep there, since I was totally exhausted. The following day, I decided to look for a job. I went to a factory where they processed wool and they gave me a job carrying wool from the first floor, where they washed it, to the third floor. And I had to carry it up a ladder in a very heavy basket. I made three trips and ran out of breath, since it had been two days since I'd eaten. I went to the office and told the foreman that I couldn't do it because the work was too heavy for me and I hadn't eaten for two days. And he told me he had no other job to offer me, but that I should go and get some food at one of the little cafés where the factory workers ate. He said to tell them I was working at the factory.

I was very discouraged, for even if I found a place where I could get my meals, the work at the factory was too heavy for me. Mr. Matthews was still at Talbotton, and I made up my mind to go there. I asked where the road to Talbotton was and got on my way. My hope was to find peaches alongside the road, or some other fruit to eat. And that's what happened. Soon after leaving, about a mile or so, I found an orchard with lots of peaches. I kept asking everyone I encountered for the road to Talbotton. That night I slept in the fields, and the following day, I got to Talbotton about eight o'clock in the evening.

I asked for the residence of Dr. F. T. Matthews and a boy was kind enough to take me to the place. The Matthews had already been notified that I had run away, but had no idea where I had gone. Mr. R. A. Matthews was not at home. He and his wife had gone to visit one of his sisters, who was married to a lawyer named Cavend[er]. I was well received at the home of Dr. F. T. Matthews. They sympathized with me for the hardships I had endured. The next morning, Mr. R. A. Matthews and his wife came to see me, and they, too, were sympathetic.

When the townspeople heard there was a Mexican boy in town, they came to see me out of curiosity and everyone was talking about "the Mexican boy." Dr. F. T. Matthews was a fine Christian gentleman and a member of the Methodist Episcopal Church South. He taught a Sunday School class and was a real gentleman in every sense of the word. Some of the people who came to see me expressed surprise and would say, "He is nearly White" while others would say, "He's as White as anybody." The doctor would reply to them, "Of course, he is White." Since there was still slavery at that time, the Whites would not associate at all with the Blacks and considered it a degradation to even sympathize with Negroes. I remember that Mrs. Matthews' daughter asked her mother one day, "Mama, is Mexican Jim sure enough White?" and her mother answered, "Daughter, James' blood is as free from Negro blood as yours or mine."[29]

CHAPTER 10

At Dr. Matthews' home, I was treated as if I were a member of the family. Every Sunday I went to Sunday school with the family, and I attended all the other church services too. The doctor was used to having family worship at home, and every morning and evening the family would meet in the parlor for this purpose. He would read the Bible and pray. In that home, I began to read the New Testament in English.

There was also a school called The Talbotton Institute which was located about half a mile from town, and through the influence of Dr. Matthews, I was admitted to it. The president of the institution was named McQueen.[30] Our own Bishop Key[31] had been a student in that school, much before I attended classes there. He told me this himself. He knew the whole Matthews family. Around this time, a preacher in charge of that circuit, named Hinton, came to town.[32] I don't remember his initials, but I got converted during a camp meeting there. I remember very well that in his prayers, he asked God for the conversion "of the Mexican boy who is among us." And I have never doubted that God answered that prayer. And I hope someday to meet him in Heaven and to put my hand in his, as I did that day when he made the call and I believed in my Savior with all my heart.

Dr. F. T. Matthews had a plantation about five miles from town, and he had about fifteen Negroes working there. That year, the plantation overseer, a White man, quit his position. (All the plantations that had slaves had White overseers.) Since school was on summer vacation, I was at home, and the doctor put me in charge as overseer of the plantation. There was an elderly Negro who supervised all the work so my only duties were to see that the workers were properly supplied with food every week and to maintain order because it was necessary to have some White person on location, to monitor what was happening. I acted in this capacity for about three months.

During the time I was there, I would go out every day with a double-barreled shotgun and walk around the fence of the field, a circumference of about two miles. I would shoot at the squirrels and kill several every day, and the Negro cook would prepare them in a stew for me. One day as I was making my usual rounds about the field, I got an urge to jump over to the inside of the fence. I leaned the shotgun against the fence. After jumping over, I grabbed the gun by the barrel and started to pull it. The trigger caught on a shrub, and the gun went off. It slipped out of my hand, and I let it go. My glove got burned, and my hat blew off. And it tore the front brim of my hat to pieces. Some of the grains of gunpowder dug into my lips, and my hand was slightly scorched. I thanked God that I didn't kill myself that day. And then I imagined how long it would have taken others to find my body had I killed myself. But God did not permit that to happen, because He had a job cut out for me. I've always believed that to be the case.

After leaving the plantation that year, it occurred to me that it would be a good idea to learn some kind of a trade. And that's what I told Dr. Matthews. Mr. R. A. Matthews had already returned to Washington. The doctor had spoken about me to a man who owned a shoe shop. His surname was Brown, and he was an elderly and very respectable man. He agreed to take me on as an apprentice and provide my food, bed, and clothing for the first year. He had no family, boarded at a hotel, and slept in the house where he had his shop. There he had a room which he used as a bedroom, and in there, he set a bed for me. As I have said, this man was very respectable, and everyone referred to him as Captain Brown. He had been a captain in the U. S. Army during the war between the United States and Mexico. He had gone all the way to Mexico City. He was a member of the Methodist Church and attended Sunday and weekday worship services with great punctuality. He treated me very well, but he had his moments of difficult temperament.

I had been working for him for about three months when one Saturday, after eating supper at the hotel, he took me to a clothing store and bought me a suit to his liking, and also a hat of the same style he wore.

It was one of those high stiff felt hats which the American boys used to refer to as "two-story hats." I tried to object to his buying that type of a hat for me, but he said, "Do you think you're better than I am?" I already knew that his word was law. The following day was Sunday, and he ordered me to get dressed for church. I always went with him. I dressed up and, instead of using the tall hat, I put on my old one. But he wanted me to put on the new one, and he scolded me, saying that I was no better than he was.

I was so embarrassed and dreaded for the boys to see me, so as soon as I saw a boy coming toward me, I would take off my hat. Since everyone knew me as "Mexican Jim," I could just imagine them saying, "Look at Mexican Jim's hat!" Mr. Brown noticed that I was ashamed of wearing the hat and when we got back home from church, he scolded me again. That night when we went back to church, I wore my old hat, but this time he said nothing. The following day I got up early and started to build a fire and prepare the shoe wax for the workers to use, as I was accustomed to doing, and he stopped me. He shouted at me angrily, "I don't need you anymore. Go home, and tell the doctor that I shall talk to him later."

I left quietly, for I knew that my words would be of no use. He spoke with the doctor later and told him he didn't want me anymore, that I would never become a shoemaker because I was too haughty.

A German by the name of Meyers who owned a shoe shop hired me to help him in whatever I could do, and I worked with him for a couple of months. By that time, Mr. R. A. Matthews found out that I had left Mr. Brown and wrote me, asking me to come to Washington, as he thought I would have a better opportunity to learn a trade there.

CHAPTER 11

When I returned to Washington, I decided tailoring would be a better trade than shoemaking, for I had had little success at shoemaking. I talked to a tailor named Mudd, and he hired me in a three-year apprenticeship, in return for room and board.

This man's family consisted of his wife, a boy about eight years old, a girl six years of age, and an unmarried lady related to his wife. They were all Catholic, but I never saw them go to church. What I did see was that they went to dances, and to the theatre, and to the circus. He always had two or three employees working for him. His wife loved to dance, and she enjoyed worldly society. She had false teeth, and I don't know how they got broken so frequently, but I was always being sent to the dentist to get them fixed. He was forever sending me on errands and, as I was already a converted evangelical, I didn't like some of their habits. Mr. Mudd would tell many lies to his clients, sometimes running them around in circles with the jobs he was entrusted to do.

I remember a young man who owned a bakery and would deliver bread on a cart. Mr. Mudd had run up quite a bill with him. One day the baker brought him a very fine cloak which he wanted altered and made into a Prince Albert frock coat. Mr. Mudd promised to do this, agreed on a price, and took his measurements. And the young man told him he wanted it for Christmas Eve. I think it was toward the end of September. The baker continued delivering milk, which was to be credited to the bill of the tailoring job he had ordered. About three weeks went by, and Mr.

Mudd had not even started the coat. One day the young man asked if he had started working on his Prince Albert coat, and Mr. Mudd said that he had and not to worry, that it would be ready when he needed it.

My ears buzzed at hearing such a shameless lie, and after the young man left, Mr. Mudd laughed about it, instead of feeling ashamed. After that, the young man would ask about his coat every day, and the standard reply was that very little work remained to be done for the coat to be ready. Starting about a week before Christmas Eve, Mr. Mudd would hide when the young man came in asking for him. His employees would lie also, saying that Mr. Mudd had gone out, knowing full well that he was in the room behind the office. This grated so much on my nerves that soon I was sorry I had ever come to work there.

On Sundays I would go visit my friends at The United States Hotel. On one of those days, they told me at the hotel that a Mexican gentleman was looking for me, and that he had left me a card with his address. That very same day I went to look for him and found him. He was a young man about twenty years of age, who was a native of my hometown, Santa Fe, New Mexico, and knew my family. He had heard people say that I was there. He was about four years older than I and had come to Washington with Don Juan Manuel Gallegos who had once been a priest in the Roman Catholic Church. I remember having met Gallegos when I was young. And he had left the priesthood, I'm not sure why. Then the people had elected him to the Congress of the United States, and he was there representing the territory.[33] The young man, José Sena, had come with him for the purpose of attending college in Washington. His father was named Juan Sena, and I remember that he ran a business in Santa Fe and was a good friend of our family.[34]

But I am forgetting Mr. Mudd and the young baker's Prince Albert coat. As I have said, every day the young man would ask for Mr. Mudd and for his coat, but Mudd would hide, and no one knew anything about the coat. Everybody in the shop knew the truth and would still lie. On the afternoon of the 24th of December, the young man came and said to the employees, "You tell Mr. Mudd that tomorrow morning I will be here for my coat, and I want it delivered without fail." The following day he came, asked for Mr. Mudd, and received the same answer as usual. And the young man said, "You tell that man that I say he is a shameless rascal. Tell him that he can keep what he owes me and also my cloak, but that he will never get to eat my bread again." As soon as the young man left, the tailor came out laughing. I was so ashamed and angry that on that very same day I left the place, and told him I no longer cared to learn his trade. And I went to The United States Hotel.

CHAPTER 12

Mr. Gallegos and José Sena had suggested that I go to Santa Fe with them as soon as the congressional term ended, but I didn't have enough money to make the trip, nor did I want to ask them for a loan. I knew that once I arrived at Santa Fe, I would have to repay it by working for them, and I did not want to return to my native land under such circumstances. But around that time, I read an ad in *The Baltimore Sun* which said, "We need fifty young men to go to New Mexico, Texas or California." I found out that these young men were needed as soldiers. I don't know how, but I got the idea to write to the Secretary of War, who at that time was Mr. Jefferson Davis.[35] When three days later I had still not received an answer, I went to his office and personally asked permission to speak to him.

He received me very courteously and motioned for me to sit down. When I brought up the subject, he replied, "Those appointments are not open yet." And I responded that I did not know that it was necessary to await an appointment in order to enlist as a soldier. Then he gave me a recommendation that stated, "If the bearer wishes to enlist, I recommend him to the nearest recruiting office which is Baltimore."

The following day I took the train to Baltimore and, as I had a copy of the paper with me, I soon found the office. There was a captain and a sergeant in the office. I handed the captain the recommendation, and he asked me how old I was. I told him I was seventeen and he replied, "You are not old enough, and besides that, you're not the minimum height that the law requires." Then I said, "Very well, good-bye," and I left the office. But I heard him tell the sergeant, “Sergeant, call that boy back. I want him for my company.”

Then he asked me if I had a guardian, and I answered that I did but that he was in Washington. He asked me if I could get his permission to enlist as a soldier. I told him I couldn’t go [to Washington] because I didn’t have enough money to pay my fare to get there. He then gave me the fare and a little note for Mr. Matthews.

That night I went to see Mr. Matthews in Washington. He had no idea what I had been doing. I handed him the captain’s note. “Why do you want to enlist as a soldier?” I told him that I wanted to return to the place of my birth, and that this would offer me an opportunity to do so. He then said, “Santiago, I don’t wish to dissuade you from your intentions. I want you to become your own man, as I’ve told you several times before.” He gave me a great deal of advice, and then sat down and wrote a note to Captain Palmer (that was his name). The following day I returned to Baltimore.

I signed up as a soldier for a five-year term on the first of April 1855. The captain said he wanted me as a bugler for his company, for which he was recruiting in order to organize a regiment. He sent me to Fort McHenry, which was located about three miles from Baltimore on the Chesapeake Bay.[36] This fort was surrounded by a lake and could be entered by land only from the side closest to the city. There were two artillery companies there. The only work assigned to me was to go out every day and learn all the military bugle calls used in the Army at the time. In a month and a half, I had learned all of the bugle calls and the military marches.

We stayed at Fort McHenry till August. My salary was paid every month. I do not remember the exact day in August when I was called to Baltimore, but I believe it was around the 10th. The following day I left with a group of recruits for Newport Barracks, Kentucky. This was across from Cincinnati, Ohio. There I stayed for about a month with nothing more to do than to go out with the drum and bugle corps and practice. It was an infantry garrison but they took us there even though we were supposed to end up in the cavalry.

During the time I was there, about ten more men signed up in Cincinnati and were sent to join our group. They were a bunch of troublemakers and wicked folks from that town. About the middle of September, we embarked on a steamship that navigated the Ohio River. That day the mob from Cincinnati really put on a show. They got drunk and wanted to fight everybody, until Lt. Low, who was in charge of our troop, ordered them jailed and put under double guard.

When we arrived at Louisville, Kentucky, and the ship was taking on passengers and the crew was loading and unloading freight, the mob, which had since been released from jail, took advantage of the occasion to start a fight with some of the folks on board. And from one minute to the next, terrible fighting broke out, and it looked like we were all in danger of being killed. The hoodlums would attack the workmen on board as well as the passengers.

In my desperate search for a hiding place, I started to take the stairway to the second floor when I saw one of my companions, Metzger, crouching under the stairway and defending himself from one of the hoodlums, named Swift. Just about this time, the captain of the ship gave orders to his crew and his laborers to suppress the tumult. As soon as the order was given, the Negroes, who were the workers on the ship, attacked the hoodlums with sticks and irons and anything they could put their hands on, and in a very short time subdued them and restored order.

Some of them tried to escape by way of the gang-plank used to load and unload freight. One of the hoodlums got half way over the gang-plank when one of the Negroes caught up with him and clubbed him with a piece of wood on the side of the head, which knocked him into the water ten feet below. The Negro dove right in after him, saying, "I'm not going to let you drown, you so-and-so." And he lifted him onto his own shoulder, and pulled him out onto dry land. The rest of the hoodlums gave up soon after this and were put in prison under guard.

CHAPTER 13

It wasn't until the end of September or the beginning of October that we arrived at Jefferson Barracks, the place where the Second Cavalry Regiment of the United States was to be organized.[37] A regiment at that time was composed of ten companies, each of which had eighty-five men, as well as a captain and two lieutenants. Each captain had worked at recruiting the men that made up his company. After a week at Jefferson Barracks, the regimental commander had all the buglers report to the field. There were twenty of us—two to each company. He spent about three hours examining us. He would put us in formation and make us play all the bugle calls and marches used in the Army. Since I had practiced much more than the others, and knew quite a bit, he named me Chief Bugler and put me in charge of all the buglers. The top staff of the regiment was made up of Col. A. S. Johnston, Lt. Col. R. E. Lee,[38] Major Hardee, and Lt. Hood, assistant. The captains of the different companies were as follows: Company A, Captain Van Dorn, Major[39]; Company B, Captain Smith; Company C, Captain James Oakes; Company D, Captain I. N. Palmer; Company E, Captain Stoneman; Company F, Captain Bradford; Company G, Captain Travis; Company H, Captain White; Company K, Captain Ives.

I do not remember the names of the rest and it is possible that I have assigned some to the wrong company but these are the names of the captains that I remember.

On the 27th of October the regiment left Jefferson Barracks and headed for the state of Texas. On this march we crossed the Indian Territory and I had the opportunity of seeing and getting to know the following tribes of Indians: the Choctaws, the Chickasaws, and the Cherokees.[40] These Indians were quite advanced in civilization. They had good farms, good ranches, good schools, and some colleges. Some of them had Negro slaves who spoke no English. It seemed very strange to me to hear Negroes speaking the language of the Indians and not understand English, since I had met literally thousands of Negroes in the South, who spoke nothing but English.

Some of the Indians I saw were quite ostentatious. I arrived one day at a house where an old Indian was sitting in a rocking chair on the front porch. I had gone there for the purpose of buying milk or eggs. He asked me to come and sit down while they brought what I needed. I was anxious to start a conversation with him and asked him what his name was. He answered, "George Washington" and these were the only words that I could drag out of him.

We traveled for about two months before reaching Texas. We crossed the Red River which marks the boundary between the Indian nation and Texas. At a small town on the Texas side of the river called Preston, there was a ferry and the whole regiment used this ferry to cross over into Texas. It took us about a week to reach Fort Belknap. I do not recall just when we arrived, but I know it was after Christmas Eve.[41] We found out they had rooms but not enough to take care of all the regiment. My company and two or three more were lucky enough to get rooms. The rest camped in tents, with which we were well equipped. A few days before arriving at Ft. Belknap, we were suddenly overtaken at night by a very strong, very cold "Norther." Many of us who had never experienced one of them learned a bitter lesson. We used to hear the soldiers say, "Texas Norther, boy."[42]

It must have been about nine at night, and since there were many fires lit across the campsite, coals and sparks were flying all over the camp. We finally had to put out all the campfires. While we were putting out the fires, the wind blew away my hat and I couldn't find it. I was suffering from a bad toothache, and that night was sheer torment. About midnight, I went to the doctor's tent with such desperation I appeared to be rabid. I stood at the entrance and started shouting for him. He asked me, "What do you want?" and I said, "I'm going crazy with this toothache!" He told me to go see the druggist.

I found the tent but he said, "Go to the doctor!" I replied, "I've been to him, and he sent me to you!"

"Well, what do you expect me to do at this late hour, in the middle of this freezing cold?"

"Do something! I'm dying!"

He got up from bed, and poured some creosote for me into a little bottle. This stopped the pain, but it burned the entire side of my mouth where the molar was, both inside and outside my mouth. I suffered from the bruise and the burn for some time, and for a long while, I couldn't even play the bugle.

It must have been in the month of April that our company and two others left Ft. Belknap, marching to Fort Mason. Fort Mason had quarters for the soldiers, but they were taken up by the two companies in our regiment which had arrived some time during the winter. At that time the captain of my company decided to make a trip to San Antonio with his family.

CHAPTER 14

There was a corporal in our company named Prusa.[43] When the captain left on his trip to San Antonio, he named Prusa in charge of an escort of five men that would accompany him. We left Fort Mason soon after the noon hour. I was a member of the escort. The soldiers had their canteens filled with whiskey when we left. That night we got to the Llano River, to the ranch of a German whose name was Martin. He was a butcher and delivered meat to the troops stationed at Fort Mason. The captain and his family stayed at Mr. Martin's home, and we camped about one hundred yards from the house.

Before we went to bed, there arose quite a dispute among the members of my group over whether we should post a sentinel or not. Some insisted it was necessary to put a sentinel on guard while others argued that there was no danger from the Indians. Among the latter group was the corporal himself. I was in favor of having a sentinel for I felt that, as soldiers, it was our duty to do so. I was more sober than the others. The discussion became so heated that they almost came to blows, and Corporal Prusa was the most agitated among us all.

As it was dark and we had no light other than the fire, we didn't see the captain, who had approached us to see what was happening. I was the first one to see him but, according to him, he had been listening to everything for some time already. As soon as I saw him, I gave the order "Attention!"

All stood at attention, as was military custom at the approach of an officer. Directing himself to the corporal, the captain said, "I have been listening to your conversation for quite a while and am sorry that after having chosen my escort so carefully, I have made a mistake. I thought that Corporal Prusa was one of the best soldiers in my company, and I am sorry to find out that he cares to post sentinel neither for his own protection from the Indians, nor for the protection of the government property under his charge." He was referring to the six saddle horses we were riding and the six mules that pulled the wagons loaded with the baggage, as well as to the forage and provisions for the trip, and the two mules that pulled the wagon in which the captain and his family rode. Then addressing himself to me he said, "Tafolla, you take charge of the escort." He then turned away and went back to the house. Since I had already told them during our dispute that I would be willing to keep the first watch, I ordered the men to go to bed.

It was about ten o'clock when I started, and at twelve o'clock I woke another man to go on duty, at two another, and at four another one, named Muth. Muth had been one of those opposed to the watch. I warned him, "Don't sit down, because if you do, you'll go to sleep." There was a great number of pigs on the ranch[44] and all night long we had been driving them away from our camp. Among the captain's belongings was a rocking chair which we had taken down. Our food supply was spread around the camp, as well as a sack of corn for the horses and the mules. Muth pulled all of these things as close to the rocking chair as possible, sat down in it, and fell asleep.

About five in the morning, I woke up and there were about fifty pigs in the camp. They had eaten all of the corn and all our food, and they had Muth, rocking chair and all, going round and round. Just about that time, a big pig got under the chair and turned it over, with Muth still in it. I chased off the pig with a stick in my hand. When he hit the ground, Muth woke up scared, asking "What's the matter?" I answered, "We're now without food for us, or forage for the animals."

When the captain came by, he discovered the news. I told him all that had happened.

"Well", he said, "let's stop at Fredericksburg to buy provisions." The only thing the pigs had left us was the coffee. The distance to Fredericksburg was about twenty-five miles, and we got as far as Cibolo [Creek] by nightfall. The coach driver, Mr. Jefferson, was also a soldier.

The next day we arrived in San Antonio. We camped in the government corral which was up against the chapel still preserved today as part of the Alamo building.[45] The area now known as Alamo Plaza was then just a chaparral of wild brush. Where the Opera building is now, and all those buildings to the North and South of the plaza, there were just little *jacales* thatched overhead with straw.[46] This was in July of 1856. The site occupied today by the Menger Hotel was a chaparral of mesquite brush. Later that same year, they began to lay the foundation for the great Menger Hotel.[47] This was the very first time I ever saw the city of San Antonio.

That same year, Captain Palmer got orders to set up Camp Verde on a creek by that name in Kerr County, eleven miles from Kerrville, the county seat, and twelve miles from Bandera Town. Both towns were made up of a few houses and a lumber mill. At this camp were built barracks for the officers and for the soldiers.[48]

It was during that time that the captain selected me to run the mail route from there to Fredericksburg and San Antonio and to the ranch of a German rancher named Zink, who lived twenty miles from Camp Verde. I ran this route every Friday. Those were dangerous times, for the Indians would come and go on their raids, stealing horses and killing everyone they encountered. I do not know why he chose me for the job. Perhaps it was because I was light on the horse. Many times, during the heat of summer, I would make the trip at night. I was always up with what was happening with the Indians. Sometimes they'd pass through before me, and sometimes after, but I never ran into them face to face.

But the captain had a reason for selecting me. One night after the six o'clock roll call, he stood before the company and said, "I want a man to deliver a letter to Castroville this very night. Who wants to go? Who knows the road?" When no one answered, I said, "I don't know the road, but I'll try it." He said, "I know you will. Get ready and come to my office for the letter."

It must have been about nine o'clock when I left Camp Verde. I arrived at Bandera about eleven that night. I woke up one of the *americano* families at that hour and told them I was a soldier from Camp Verde taking a letter to Castroville. The man came out with a rifle in his hands, for in those days, no one could leave his house unarmed for fear of the Indians. And after he was convinced that I was acting in good faith, he gave me the directions I needed.

CHAPTER 15

I went across the Medina River at this point. About ten miles from there, I crossed the Medina River again, and about three miles from there, came to a place they called Mormon Town. There was a colony of some ten or twelve Mormon families in the village. It must have been about 1:30 A.M. when I arrived, and I headed in the direction of a house that was all lit up.

As I came closer, I heard music and saw that the people were dancing. I went up to about ten paces from the house, called out "Hello!" and the music stopped immediately. Some half a dozen men came out, holding firearms. After meeting me and learning the purpose of my trip, they invited me into the home and started dancing again.

There were about fifteen women and girls there, and another handful of men, and everyone appeared to be joyful and happy. They invited me to be seated, and an old gentleman who appeared to be the leader sat next to me and said, "We're celebrating a wedding that just took place in our families."[49] He welcomed me and asked me if I wished to dance. I thanked him and told him that I had to be in Castroville at the break of dawn. He ordered someone to feed corn to my horse, and took me to another room where there were tables loaded with all kinds of the most delicious foods imaginable. I did not see any liquor of any kind, nor signs that anyone had been drinking. That night shaped my opinion that the Mormons were good people. I stayed there about an hour. The old gentleman who acted as my host invited me to stop by on my way back.

I arrived at Castroville about six o'clock in the morning. The letter I carried was for a Mr. Vance. I asked for directions to his home, and soon found it. But I was told there that he was in San Antonio, and would not be back till after the noon hour. I had no money on me for food for me or for the horse. I tied the saddled horse to a large oak tree and went to walk around the town. By then, I was very hungry.

About one o'clock, Mr. Vance arrived and gave me a letter in answer to the message I had brought him. Then I told him that I had not eaten and my horse had not been fed either. I told him that I had no money, but that he could charge the expense to the government, as I was on government business. He then told me that the letter I had brought him was in regard to a contract with the government about a shipment of corn which would soon arrive at Camp Verde. He ordered food for me and for my horse.

It was four o'clock when I left town to return to Camp Verde. About nine o'clock at night, I passed through Mormon Town and everything seemed very quiet. Between Mormon Town and Bandera I stopped and slept for about three hours. When I got back to Camp Verde the *carretas* [big wheeled carts] were already unloading the corn. The captain could have saved me from that trip but since he had the authority to do so, he could send a soldier anywhere he wanted to. Now I get the full meaning of the passage of scripture where the centurion tells the Master about his authority (Mat. 8:9).[50]

On another occasion the captain sent me on a hurried mission to Fredericksburg to get a doctor. On still another occasion, although I had never been over the road before, he sent me after the Justice of the Peace at Kerrville to bring him to the Guadalupe River near Kerrville to perform an inquest and take care of the body of a man who'd been killed about a mile from Camp Verde. This is how he always used me on similar occasions. I believe he did it because I was *mexicano.*

On one of these trips to deliver a message, I was caught unprepared by a cold "Norther," accompanied by rain and freezing rain. I got back to camp about six o'clock in the afternoon. I lived in the quarters assigned to sergeants and corporals. When I arrived, there were six men in the room and the fire had not been started. I went out, got wood, and built a fire. Some of them remarked that if I hadn't arrived, they would have frozen to death.

I had washed my face and was standing right in front of the mirror combing my hair and saying that all of them were lazy, and that was the reason the fire had not been started. A fellow named Dennis who was a corporal told me to shut up or he would kick me out of the room. I told him he wasn't man enough to do it and that if he even tried, I'd shoot him.

I was very angry because they were making fun of me, after I had built the fire for them. Just then, Dennis came at me intending to deal me a blow from behind, but I saw his movements in the mirror. I drew my gun out of its holster immediately, fully intending to shoot him, but all the others jumped in and held me. They took me to the guard house and jailed me and reported the incident to the captain. That night I slept in prison. The pay master had come that very day. I was hoping that the captain would let me out on the following day when the shift changed on guard duty, as was customary to do with the prisoners who were not criminals. But he didn't do it, and instead had me report to his quarters, to talk to me.

He told me that he would let me out if I apologized to Corporal Dennis. I told him that I felt I had a right to defend myself, and that I had done it in self defense. I realized that what he wanted was for me to humble myself, and this is exactly what I should have done.

Downtown San Antonio 1854, looking west on Commerce St. (UTSA's Institute of Texan Cultures.)

CHAPTER 16

But I was very proud. I had lost my religion, and Satan had taken advantage of this occasion. The captain said to me, "If you don't apologize to Corporal Dennis, you'll have to be a prisoner for quite a while and finally be court-martialed by the military." I said to him, "If you keep me in prison for the rest of my enlistment period, which is three-and-one-half years, I will not apologize to Corporal Dennis."

There was a pit about thirty feet deep in the center of the camp which covered an area two hundred yards square. On one side of the square plot were the stables for the horses. On another side were the stables for the camels,[51] while the third side was taken up by the mules. On the fourth side of the square were the barracks and the commissary. In the middle of the square was the large pit which was started as a well and never finished. It was about thirty feet deep. Soldiers who got drunk were imprisoned in this place to make sure they couldn't get any liquor while held there.

But they [the drunk soldiers] had devised a way to get it. They made an arrangement with some of those who had already been released to bring wine to them at night. The one who brought the wine would whistle and drop a little rock into the pit. The ones below stretched out a blanket between them and held it high over their heads. The one above would then drop the bottle or the canteen into the well. As it was night, I landed in this place and the next day the captain had me pulled out to report to him. He said, "Tafolla, I don't want to see you in the pit with the drunks. And I don't think you want to be there either. I want you to promise me that you will not go looking for a fight with Corporal Dennis, and I will set you free." I had [earlier] told the captain that the very day I was freed, I would settle the score with Dennis. I told him I couldn't make that promise. He then sent me to the guard house, but not to the pit.

Since the soldiers had recently been paid, they spent their time gambling trying to get each others' money. I was the only prisoner in the guard house. The others were in the pit. It was beginning to get dark, and the corporal in charge could not find the sentinels to take the first shift which began at 6 p.m. They were all busy gambling in the quarters.

Corporal Jim said to me: "Jim, keep an eye on these horses while I go to look for the sentinel." At that very moment, the idea of getting on a horse and deserting entered my mind. I picked out a horse and took him beyond a creek that was about three hundred yards from camp. I tied him and came back to get my blankets from the guard house. It was winter and I knew I had to travel all night. When I returned to the place where I had tied the horse, I found out that he had gotten loose and was nowhere to be seen. I left my blankets there and went back to the corral to get another horse. I was untying him when I heard Sergeant Brown dragging his sword and coming my way. Right away I started talking to the horses as if I were trying to quiet them down. Just about that time, the horse I had left beyond the creek got back. The corporal asked me what had happened. I told him that I was tying up a horse that had gotten loose. He asked me to please watch them a little longer, and he went to look for his men. By that time it was very dark. I took the horse, laid the blankets on him, and mounted him. I traveled all night long, without knowing the way, but I had an idea of the general vicinity.

About four o'clock in the morning I got to the ranch of Don Manuel Hernández on the San Geronimo Creek. Federico Rivas, who had known me since the time we were at Belknap, was there. He had seen me as a prisoner the day before at Camp Verde. As soon as he saw me, he realized that I had deserted. I told him that my plan was not to desert, but to appear before the colonel of the regiment and ask that justice be done in my case. My friend loaned me a saddle and I arrived at San Antonio about eleven that morning.

Straightaway, I went to the office of the colonel and related to him all the circumstances regarding my case. "Tafolla," he said, "I'm sorry you got into this trouble. Take your horse to the government stables, and go enjoy the town. I think they'll come after you. This very day, I'll write Captain Palmer and tell him you've presented yourself before me here."

Two days later a guard by the name of Jim Claude and Sergeant Hemphill and two soldiers came to the office of the colonel, and learned that I was in town. There was a certain hotel that catered to the military, and they went to register there. I was eating supper when they came in and, as we were good pals in the army, they greeted me cordially. Two days later we started back to Camp Verde.

When we arrived, the sergeant went to give his report to the captain, who ordered me to the guard house. That night they put me back in the pit. The following day the captain had me report to his office. "Tafolla," he said. "I have done everything I could to save you from punishment, and as you well know, I have always had more consideration for you than for any member of my company,[52] and instead of taking my advice, you have made yourself liable to criminal prosecution. The punishment for a deserter is very severe as you know but now I am going to give you another opportunity. If you will promise that you will not try to renew the fight with Dennis in the barracks or on the grounds of Camp Verde, I will speak on your behalf to the military authorities. And if upon some occasion, you and Sergeant Dennis meet outside this camp and want to settle affairs, you may do so, but not in the quarters or within the limits of this military post. Because, as you know, my company has always enjoyed a good reputation, and I don't believe you want to be the cause of our losing it." Truthfully, our company had always had the best reputation of any of the companies in our regiment. I asked for the captain's forgiveness, for having been so stupid, and promised to abide by what he had asked. That same day he set me free.

CHAPTER 17

About eight days later, it became necessary to send a detachment of soldiers to catch up with some Indian raiders who had stolen the horses from a Dr. Nowlin[53] who lived on the Guadalupe River about thirty miles downstream from Camp Verde.[54] I enjoyed these expeditions and asked the captain permission to go. My pistol had been taken from me on the day I had the trouble with Dennis. They gave me another one, which was in very bad shape, and a rifle of the kind the army used at that time.

As a guide, we had J. P. Rodríguez who was quite an expert in trailing Indians.[55] At the headwaters of the Guadalupe River, we found the Indians' trail. The guide was riding a government mule that he always used when trailing Indians, and the mule was as skilled as he at following the trail. He was very experienced at trailing Indians. He would put the mule on the trail, and he'd be looking ten to twenty yards[56] ahead. We didn't see anything, but he was so well-versed in trailing that he'd look ahead to the grass or the weeds that were folded over or the small rocks that had been moved, and he kept us going at a fast trot.

Around six days later, we came within about six miles of Fort McCavitt, which is located at the head of the San Saba River. Since some of our horses had lost their horseshoes and were limping from the rocky terrain we'd covered, the sergeant in charge of our detachment decided that we should go there and have the horses shod. We got to the fort about one in the afternoon and, after replenishing our supplies, left at five the same day. Just about dusk, we got back to the place where we had left the trail and camped there for the night.

I have never been superstitious, but that night I dreamed that we had caught up with the Indians and that we had engaged them in a skirmish. I dreamed that one of the Indians had cut off from his companions and that John Martin and I had followed him. We'd caught up with him in a thicket, wounded him, tied his hands, and had him going ahead of us as a prisoner. I dreamed also that all of a sudden, with his hands still tied, he'd snatched a dagger from Martin's cartridge belt and had plunged it into Martin's body, close to his heart. In an instant, without giving me time to defend myself, he had plunged the dagger into my groin just below my cartridge belt. I felt the pain, cried out, and woke up. Martin, who was sleeping next to me and heard my cry, woke up, and asked me what had happened. I told him all about the dream, and he laughed. Dawn was breaking, so we got up, fixed our coffee, and got back on the trail.

About four that afternoon, one of the spies who flanked our right came to tell us that the Indians were camped close by. About the same time, the soldiers in the lead got a glimpse of them. It was not the same band of Indians we had been following. It was a different group, coming to steal horses and kill whoever might try to stop them, as they always did. There were only about fifteen of them, and they were eating. They had only three beasts of burden, a mule and two horses, on whose backs they had all their provisions loaded. There were cured buffalo hides, leather moccasins, chamois cloths, and we could tell they were on an expedition by what they had packed.

The animals had been turned loose to graze on a plain to our right, and one of their men had gone out to round them up. When the Indian coming back with the animals saw us, he sounded an alarm, left the animals, and ran to join his companions. To the left of their camp there was a thicket of small trees and thick brush, and through it ran a creek fed with water from a place the English-speakers call Kickapoo Springs.

The Indians went into this thicket and began to attack us with arrows and firearms, and they were well armed. Our arrival had surprised them so thoroughly that some of them had left their guns next to the fire where they'd been eating. One of them came out of the woods to get the rifle[57] he'd left behind, and just before he got back into the woods, he was killed by a bullet from our men, who had dismounted and begun to fire.

The Indians' intention was to take our horses, and they had divided up into three bands. Some were attacking us on the right, some on the left, and some from the front. And since they were hidden by the thick brush, we couldn't see them until they charged us. At our first attack on them, our guide, who was the only one who'd ever fought Indians before, had said to the sergeant, "Don't let the men go into the thicket on horseback, because not a single one will come out alive." The sergeant then gave orders to the men to dismount from their horses and fight on foot. He ordered us to tie the horses as the army had trained us to do, so that one soldier could guard twenty horses.

I dismounted and, since my horse was very tame, I just dropped the reins down and left him loose. The Indians made several attempts to get between us and our horses, so that they could take them. On one of those attempts, a number of the Indians had lined up on my side, determined to cut us off from the horses, while other Indians were making a similar attempt on the right flank. On the left my companions and I were lined up, ready to shoot.

Then I heard the guide's voice calling to the sergeant, "Everyone get on those horses, or we'll lose them!" I thought everyone would do as he said. I turned and saw that my horse had followed me and was close by. I threw away the rifle I was fighting with, mounted my horse, drew my pistol, and went to meet the Indians who were riding toward us single file. The Indian at the head of the line adjusted his shield, and waited for me. He began to jump up and down and shout a war whoop, as they were accustomed to doing before fighting. He waited for me to come to him, while the other Indians went back.

Policarpio Rodriguez, Santiago's friend, cousin, and fellow soldier in the Second Cavalry, with his grandson, John Hill, c. 1890.

CHAPTER 18

I fired the first shot at a distance of about twenty yards, but I was at a gallop and coming up on him quickly. I fired the second shot at him when I was about ten feet away, and I thought the third would hit him right in the head. However, as I pulled back the trigger, the shell of the bullet I'd fired lodged in the cylinder, and before I knew it, I was right upon him. As he kept protecting himself with the shield, not a single bullet hit him, so what I did was to club him on the head with my pistol. At that moment, I felt an arrow penetrate below the waist and felt the same pain I had experienced in my dream the night before. The Indian straightened up, fully expecting me to fall from the horse. When he saw that I was not falling, his expression changed to one of surprise, and he threw his shield over his back, and retreated to the thicket.

I started to go after the Indian, but saw that no one else had mounted his horse. One of my good friends, Fred Metzger (who now lives in Hondo, Texas)[58] had followed me on foot and had picked up the gun I dropped when I hit the Indian. Right then, I saw an Indian shoot John Martin with a rifle, and the bullet went right through him, piercing his body just below the left nipple and crossing through his lungs.

All this made me think about the dream. The Indian hadn't killed me because I had taken off my heavy military overcoat, rolled it up very tightly, and tied it to the horn of my saddle with a leather strap which I kept for this purpose. The arrow had pierced about sixty folds of the overcoat, including the sleeves, the very thick lining, and the thickest part of the coat, the back. It also pierced my cartridge bolt, my belt, my clothing, and about half an inch into my body. All this was due to the

Providence of God, who saved my life that day. I believe that if the Indian had killed me, he would have mounted my horse and succeeded in stealing all our horses.

When we attacked the Indians, our guide, J. P. Rodríguez, set out first of all to herd off the Indian horses and put them behind ours. He stayed on guard, watching the horses and watching the fight. He said he could see the dust come off the Indian's shield every time one of my bullets hit. He wrote about all of this and about this skirmish with the Indians in his book about his life.[59]

I remember that the second lieutenant of our company was killed in like manner by the Indians on the expedition to the Wichita Mountains. There were four companies under the command of Major Van Dorn. Lt. Van Camp met an Indian under the same circumstances as I had, and the Indian pierced his body with an arrow and killed him. He fell off his horse, and the horse ran off alone in the direction of the Indian camp. I knew that horse, and he was spirited, and I don't believe the Indians were ever able to catch him.

After the Indian wounded me, I dismounted and began shooting with my rifle again. Two of our men had taken our wounded man to the back. At that moment, a very tall Indian came toward me, hiding behind the brush, and he was holding what we generally called a "dragoon pistol"[60] in his hand, looking for an opening to take a shot at me while I was loading my gun and getting ready to shoot at him. But Fred Metzger was also hiding in the brush, and the Indian couldn't see him even though he was only about fifteen feet away.

Just at the moment the Indian raised his rifle to shoot me, Metzger shot him in the stomach. The bullet must have pierced his body and broken his back bone because the Indian jumped up and fell dead on the spot. I was taking aim when this happened and I said to Metzger, "You stole my shot." He replied, "I got him." As soon as that Indian fell, two others came out of the woods and dragged him away. They sounded a retreat[61] and went away crying.

The sergeant wanted us to go back into the woods to find out how many we had killed, but J. P. Rodríguez advised against doing this. He said the wounded Indians in the thicket would kill our men, and it was already getting late.

One of our men had lagged behind in order to avoid the fighting. The sergeant laid him low and called him a big coward. This was the very one who, back at the soldiers' quarters, while drinking whiskey, always acted very brave and intimidated others. The sergeant charged him with cowardice before the enemy and he was put in the guard house. Later he was court-martialed and discharged from the service with a "dishonorable discharge, unfit for the service of his country."

CHAPTER 19

As soon as the Indians left, the sergeant made McDonald carry all the Indians' provisions and bring it to where the pack mules were. I must say here that Dr. Nowlin and another American by the name of Moore had come along with us in this campaign and when the fighting began, the doctor got off his horse and started firing a double-barreled firearm that he carried, and then took out his pistol and kept on firing till his ammunition gave out. Then he went back to where his horse was and began to load his guns while hiding behind the horse. He was a tall man and was wearing some strips of a red blanket material around his legs. Perhaps the Indians thought that he was the captain. A rifle bullet intended for him hit his horse behind the ear and killed him. The doctor immediately began to remove the horse's saddle and pulled it back, muttering, "That's the way these cowardly rascals act. They get into a chaparral instead of coming out in the open plain and fighting like a man."

Then we had the providence to leave that place, to take the wounded man to the doctor at Fort McCavitt, the nearest place [with medical facilities.] We saddled my horse with one of the Indians' saddles. They improvised stirrups for us and stretched buffalo hides to hold both of us wounded men on the same horse.[62] We left the place at dusk. I asked the sergeant if we could get to Fort McCavitt that night and he answered that he thought we'd get there by daybreak. We traveled in the direction the guide took us.

About midnight, some of the soldiers began to get rebellious, saying that we were lost and that if we kept going in that direction, we would never get to Fort McCavitt. The sergeant stopped and consulted the guide. The guide told him, "We're not lost. Pretty soon we'll get to the lake where we got water this morning."

We had traveled about a mile when I heard the squawking of ducks, and the guide said, "Listen, those are the ducks on the lake." As a matter of fact, we soon arrived at the lake, drank water, and continued our march. Every little while, the soldiers would ask the guide, "How far is it to Fort McCavitt, Polly?" and he would answer, "It's not very far now." It must have been five o'clock in the morning when we heard a voice shout, "Who comes there?" It was the sentry from Fort McCavitt.

As soon as we arrived, we reported to the commander, and the soldiers in the fort got up and greeted us cordially. They took the wounded man and me to the hospital about six o'clock that morning. The doctor came and took charge of the wounded man. The wound on my groin was a little swollen and it was hurting, so I chose to stay in the hospital. The doctor applied a pomade and dressed the wound. There were stories around at the time that the Indians used poisoned arrows and that is what I feared. However, the fact that the arrow had to cut through so much material must have removed much of the poison, and I believe that's what happened. Twenty-four hours after our arrival, Martin died. We buried him at three o'clock that afternoon, with military honors. All the troops and the military band accompanied us. The next day we started back to Camp Verde.

When we arrived at Camp Verde the captain said to me, "Tafolla, you barely escaped being killed by an Indian." I replied, "If I'd had my own pistol, I would have killed the Indian." I always kept my pistol clean and in good condition, but it had been taken from me when I fought with Dennis, and the one returned to me was another.

In 1858 we began to hear rumors of a war against the Mormons in Utah, and our regiment marched out to join the companies stationed at Camp Cooper.[63] But a few days after our arrival, we were notified that everything had been settled. Two other companies and ours remained at this place for about two years.

It was during this time that the expedition to the Wichita Mountains was organized,[64] and our company's second lieutenant was assigned to it. About a mile from the military stockade was what was called "the Indian Reservation," which consisted of a tribe of Comanche Indians. There were about a thousand of them, including their families, who were under government protection. They had made a peace treaty, and the United

States supported them and established schools for their children. Their chief's name was Ketomsey.[65]

These Indians had kinfolks and friends who were among the wild Indians, who refused to cede their allegiance to the government. They would enter the settlements at Cooper, steal horses, and kill people. On many occasions, the settlers would follow their trail right up to the Indian reservation, and want to kill the Indians there, thinking they were the ones at fault, while the thieves would get away with the crime.

The commander of the troops would tell the people to point out the thieves and give them up, but they could never find either the thieves or the goods. One time, the Indians caught two of the wild Indians who'd come to plunder, and they turned them over to the troops. They said that these Indians belonged to Buffalo Hump, and that he was camped somewhere in the caverns of the Brazos river.[66]

The colonel organized an expedition of two hundred men, and he and Major Thomas left to find Buffalo Hump. We took the two Indians with us as prisoners, and they were going to lead us right to the place. Chief Ketomsey, the chief of the reservation, and twenty five friendly Indians were also in our party. The Indian prisoners told us we'd be there in five days, but we traveled ten days and found nothing. Every day they'd tell us we would get there the following day. At last the colonel realized we were being tricked. Ketomsey would admonish the Indians with him for an hour every night.

Among the Indians, there was a Mexican captive named Encarnación. He belonged to Ketomsey and would tell me that his chief's lectures were counseling the Indians to be loyal to the U.S. government and advising them not to become involved should the troops succeed in finding and engaging the Buffalo Hump Indians in a skirmish. He told the Indians that the troops were not there to fight but to persuade the Indians to make peace with the government.

CHAPTER 20

There were six companies on this expedition. Every day, as we began to march, one company would travel about a mile to the right of the main ranks, another company about a mile to the left, and two in the middle. Behind these, came the pack mules. The Indians would go about a mile or so ahead, as explorers. Any indication of danger was reported to the company at the center. One day the Indians reported to the captain of my company (which happened to be on the left) that they had discovered fresh horse tracks. The colonel was notified and had the troops stand ready for action while the Indians examined the tracks more carefully. Soon after this, one of the Indians came to report that the tracks were made by two horses which were running loose and that they were going to lasso them.

Soon we saw the Indians come up at a fast gallop, with two horses in front of them. They passed right by our company. They were Mexican ponies, and very well fed. They went past twice. The soldiers couldn't lasso them because there was a strict order that no one was allowed to travel on his horse faster than a trot. Anyone violating this regulation during the day would have to give up his mount and walk for the rest of the day. I was aching to do it, and would have given my right arm to rope one of the ponies, but I dared not disobey orders.

On one occasion, when the Indians attacked the company behind the horses, we saw two loose horses. My lieutenant dug the spurs into his horse to follow them. I already had my lasso made and ready, and I followed him. The sergeant shouted "Halt! Halt!" but I didn't stop. As I was riding a fast horse, I caught up with the two horses in just a moment.

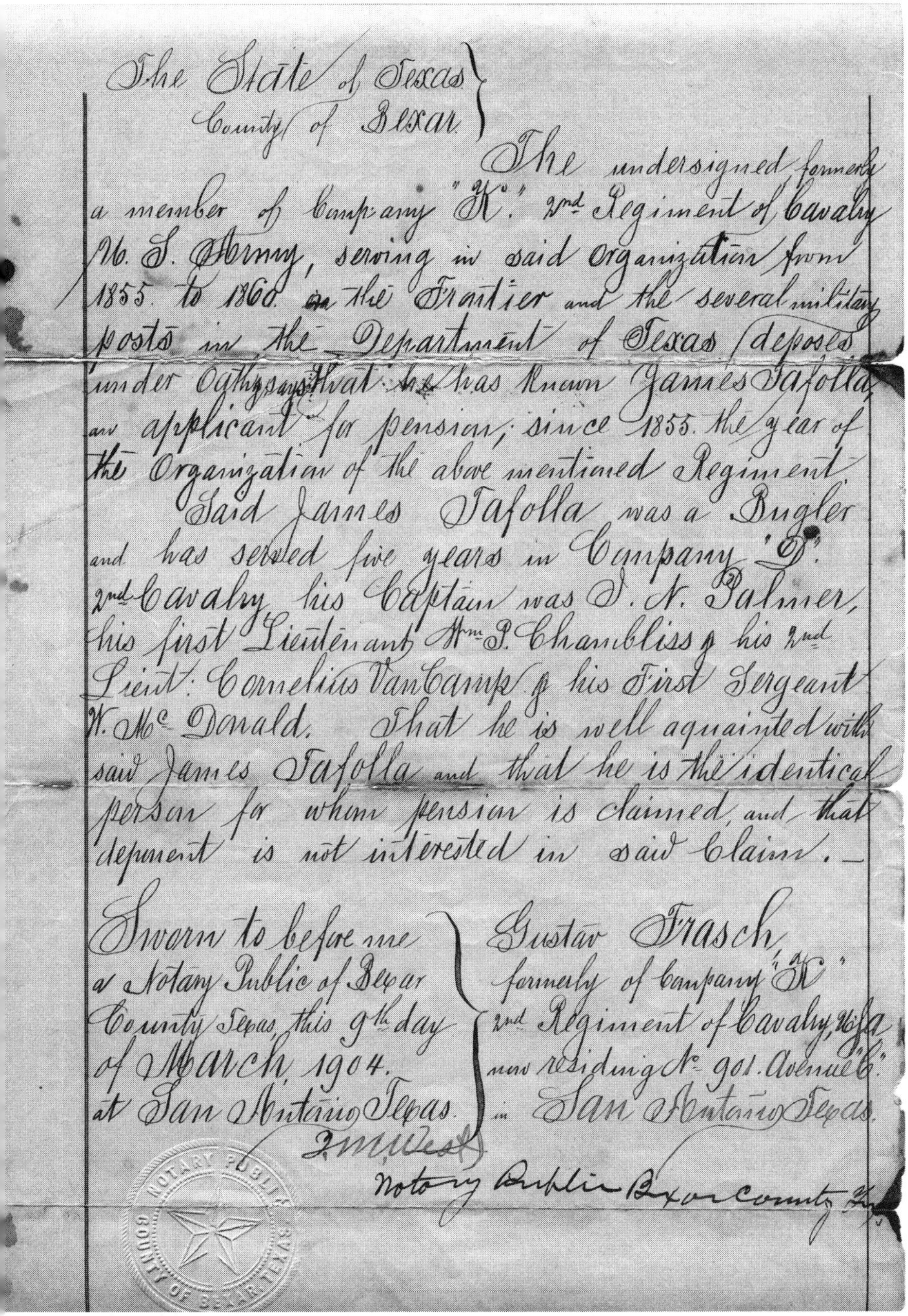

The State of Texas }
County of Bexar }

The undersigned formerly a member of Company "K" 2nd Regiment of Cavalry U. S. Army, serving in said Organization from 1855 to 1860 on the Frontier and the several military posts in the Department of Texas deposes under Oath that he has known James Tafolla, an applicant for pension, since 1855 the year of the Organization of the above mentioned Regiment.

Said James Tafolla was a Bugler and has served five years in Company "D" 2nd Cavalry his Captain was I. N. Palmer, his first Lieutenant Wm. P. Chambliss & his 2nd Lieut. Cornelius Van Camp & his First Sergeant W. McDonald. That he is well aquainted with said James Tafolla and that he is the identical person for whom pension is claimed, and that deponent is not interested in said Claim.—

Sworn to before me a Notary Public of Bexar County Texas, this 9th day of March, 1904. at San Antonio Texas.

Gustav Frasch, formerly of Company "K" 2nd Regiment of Cavalry, U.S.A. now residing No. 901. Avenue "C" in San Antonio Texas.

F. M. West
Notary Public Bexar County Tex.

This letter was written by Gustav Frasch to attest to Santiago's participation in the U.S. Army Second Calvary. (Courtesy of C. Tafolla.)

They were running so close to each other that it looked like they were tied together. I threw the lasso on both of them, and the moment they felt the rope about their necks, they stopped. They were tame horses that the Indians had tired out and turned loose, no telling how long ago. The lieutenant rode up and said, "Give me one, Tafolla." And I gave him one right away, for I was scared of the punishment, but no one ever said anything to me about it. Three days later, we found two more, a stallion and a mare. Perhaps the Indians had ridden them hard.

Finally, we arrived at a very large pond and made camp for the night. Each company unsaddled and set its horses loose in separate areas, as we always did. I had loaned my horse to one of my companions. He had somehow tired out the horse, although he was among the men in the rear guard who rode behind the pack mules. Those horses had gotten away from the Indians, but the Indians hadn't followed them, because their own horses were tired. When my horse got back, I knocked my friend's saddle off of him, and jumped on bareback in the direction the Indian horses had gone.

After a mile or so, I saw them standing there. I got off my horse and tied him, and he neighed at the other horses. I dropped down in the tall grass. Pretty soon the other horses came to join up with my horse. I had a lasso ready in my hand. I crawled quietly to where my horse was, and stood up, using him as a blind. I tried to whip him gently to get him closer to the other horses, but he kicked me in the stomach and knocked me down. The huge commotion I made scared the other horses off, and they ran about another quarter mile away.

For a long while, I rolled on the ground, groaning with pain, but as soon as the pain lessened, I got on my horse again and drove the horses back to camp. I asked two of the fellows to stretch a long rope on the ground between them and to come up with it as soon as I drove the horses across the rope into our own herd. When they did this, the stallion stopped, but the mare jumped over the rope and went toward the next company's herd. The men in that company tried to rope her, but I jumped bareback on the stallion, muzzled him with my rope, and followed the mare.

The soldiers in that company shouted at me, "This one's ours!" but when the mare saw the stallion, she came right up to us, and I said, "Well then, take her." I put the stallion at a gallop, and the mare caught up with me and stuck right by my side. I slipped the rope over her with no difficulty. I gave the stallion to the fellows who helped me catch him, and sold the mare that same day for sixty dollars.[67]

CHAPTER 21

We had just about used up all our provisions by this time,[68] but, as there were plenty of buffalo and deer around, we didn't go hungry. And we never did find any signs of Buffalo Hump. At night, our captains met with the Indian Chief to determine what to do with the Indian prisoners. The sentence they imposed on them for being such liars was to cut off their braids, shave their heads, and turn them loose without firearms. The Mexican captive confided to mc that this was thc most dcgrading punishment that could be given to an Indian. We had already reached the Staked Plains near New Mexico when we decided to turn back. The following day the Indians all went back to Ft. Cooper.

On our way back, we had stopped to camp very early one day when I asked the captain for permission to go and kill a buffalo. I found a herd a mile away from camp. I shot and wounded a large bull, but I wanted to have the satisfaction of killing him, so I followed him. He broke off from the herd and skirted some hills. I stopped to load my pistol, and then followed him. The terrain was rough and covered with ravines. I was riding fast when the horse tried to jump over a ravine and fell into it, failing to make the bank. About ten feet away, at the head of the ravine, I saw a puma.[69] He was startled, and maybe scared by the noise of the buffalo herd which had just thundered past, or by the hoof beats of my horse. The puma looked like he was ready to pounce on me. My horse had landed on his knees, but I wasn't about to let go of him. The horse got up fast, and I got out of there, trembling with fear. I forgot all about the buffalo bull and have no idea what direction he went. It was late at night when I got back to camp, and everyone was worried about me. They thought I'd been killed by the Indians.

There was another time, too, when I shot a buffalo and followed him. He jumped off a precipice and broke his shoulder blade. I went down to the creek bed where he was lying and got very close to him with my pistol in my hand. At a distance of about ten feet, I took aim and hit him on the forehead, right between the eyes. But every time I shot, he just shook his head at me. On this occasion, a friend of mine had followed me. Jim Claude was one of the government guides, and had come looking for me when he heard the shots. As I shot my last bullet, the buffalo bull got up and charged me. I tried to escape, but my spurs got tangled in the underbrush and I fell backwards. And since the bull's shoulder blade was broken, he fell too, toward me, and his big snout touched my feet. At this instant, Jim Claude arrived and shot him in the eye with his rifle. I didn't know they had so much padding on their forehead; no wonder my bullets didn't even penetrate.

It was about one month after being at Camp Cooper that the Wichita Mountain expedition was organized, and that Major Van Dorn's troops fought and destroyed the Indian forces. This happened one night just as it was dawning. Notices of this defeat reached the Indian reservation that evening. About midnight, the Indians raised such a clamor that it could be heard at our camp a mile away. Some of us went to investigate and we found the whole camp awake and walking around, the women cutting their breasts with knives, and the men mourning and singing war songs. We knew this because of what their captive told us. Later that day, we received notices about the encounter and the battle, and the death of Lieutenant Van Camp. This all happened in 1859. Soon after Van Camp's death, another second lieutenant named Cunningham was sent to replace him. Since the captain and the lieutenant named Chambers were absent, Sergeant Cunningham assumed full command of the company for a while.

On the last day of March 1860, I completed five years of military service in the army of the United States of America and on the first of April, 1860, I received an Honorable Discharge.

From there, I came to San Antonio with the intention of marrying a girl who was a relative of J. P. Rodríguez. I'd met her at Polly's house, and had promised to marry her as soon as I left the service. The wedding took place shortly after my arrival in San Antonio, Texas.[70]

Soon after leaving the service, McDonald, a staff sergeant in my company, also completed his term of enlistment, and went to Bandera County. He bought fifty cows, with the intention of starting a cattle ranch. He contracted me to take care of them on a "share" basis for five years.

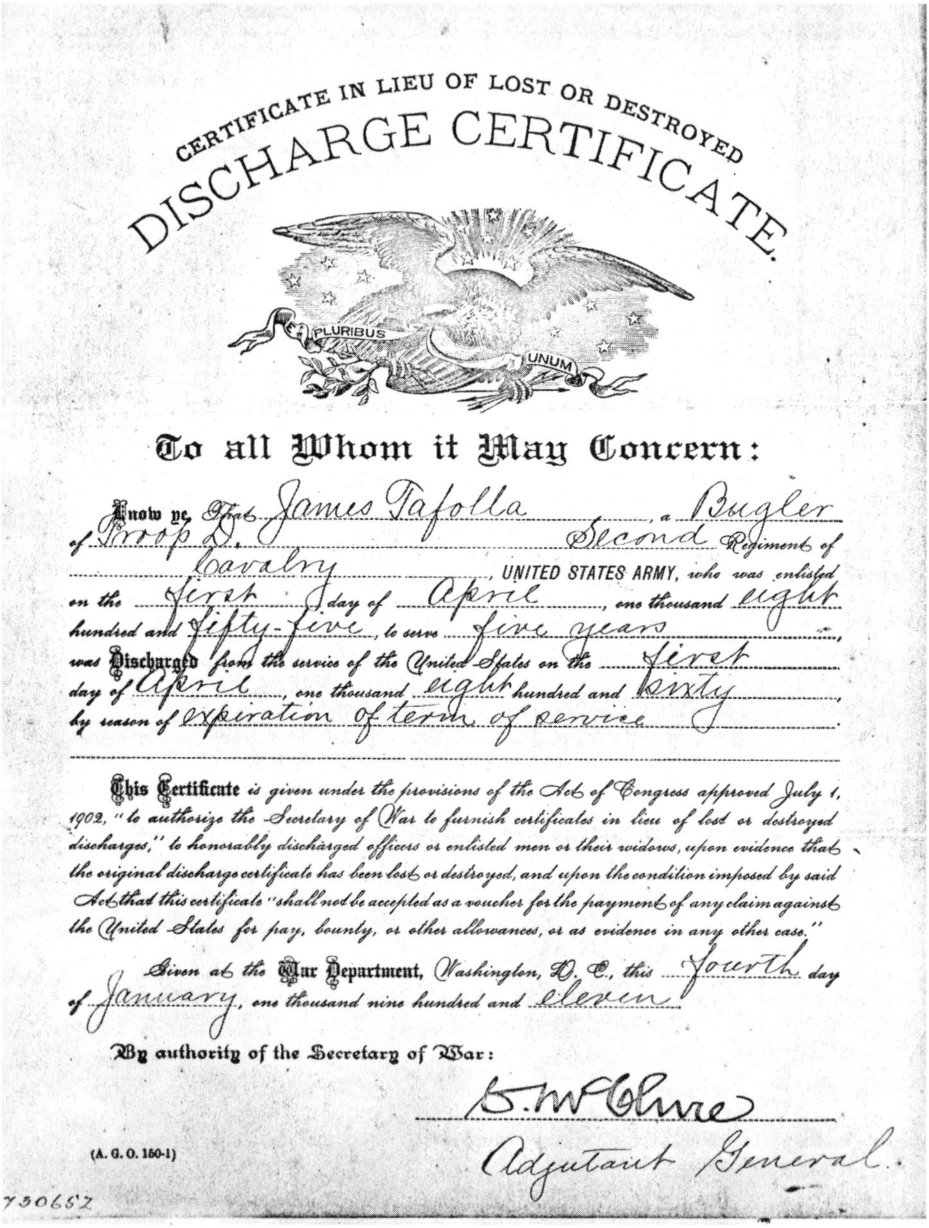

CERTIFICATE IN LIEU OF LOST OR DESTROYED
DISCHARGE CERTIFICATE

E PLURIBUS UNUM

To all Whom it May Concern:

Know ye, That James Tafolla, a Bugler of Troop D, Second Regiment of Cavalry, UNITED STATES ARMY, who was enlisted on the first day of April, one thousand eight hundred and fifty-five, to serve five years, was Discharged from the service of the United States on the first day of April, one thousand eight hundred and sixty by reason of expiration of term of service.

This Certificate is given under the provisions of the Act of Congress approved July 1, 1902, "to authorize the Secretary of War to furnish certificates in lieu of lost or destroyed discharges," to honorably discharged officers or enlisted men or their widows, upon evidence that the original discharge certificate has been lost or destroyed, and upon the condition imposed by said Act that this certificate "shall not be accepted as a voucher for the payment of any claim against the United States for pay, bounty, or other allowances, or as evidence in any other case."

Given at the War Department, Washington, D. C., this fourth day of January, one thousand nine hundred and eleven.

By authority of the Secretary of War:

B. W. Clure
Adjutant General.

(A. G. O. 150-1)

730652

Santiago Tafolla's Discharge Papers from the U.S. Army, Second Cavalry, April 1, 1860. (Courtesy of C. Tafolla.)

CHAPTER 22

In that same year, the war broke out between the United States and the Confederate States, and when the army was set to march out of the state of Texas, Sergeant McDonald deserted, came to where I was, and sold the cows. He promised to pay me $250.00 for all of the cattle. When I delivered the cattle to him, he asked me to accompany him to San Antonio, where he would withdraw the money from the bank. We arrived in San Antonio, and I stayed at the house of some of my wife's relatives. The sergeant had told me to meet him at the hotel first thing in the morning, and he would pay me. The following day, when I went to get paid, I was informed that he had taken the stagecoach that night to Indianola. I never saw him again. He left me two horses that might have been worth about $50.00 each.

That year I got 160 acres of land, which was the amount the state of Texas was granting to every citizen who was a head of household. I built a ranch on a particular branch of Privilege Creek which was called Bear Creek. I lived there for some time with my brothers-in-law, who'd come to live with us there in Bandera County. J. P. Rodríguez had established a ranch on the main branch of Privilege Creek approximately two miles from mine.[71]

The Indians bothered the settlers on the frontier a lot. There was a man there at that time named John James, who owned a lot of acreage. He sold me a piece of land—three hundred acres[72]—close to Bandera City. I intended to start a ranch there. This man had advertised in the newspapers that he owned land in all the frontier counties of Texas and would sell it at a dollar an acre, in Confederate money. I paid him $150.00 and promised to pay him the balance later, with interest. This was at the beginning of the war. About three years later, I went to pay the balance, and he wouldn't take it, because Confederate money wasn't worth but fifty cents on the dollar by that time. So I lost it. I lost my money and the land.

In 1862 I got a job at Camp Verde taking care of the camels the government had there, and I worked at it for about eight months. In the fall of that year I went to San Antonio and joined the Confederate Army, under the battalion of Col. Duff.[73] During that winter, the battalion guarded the border from Laredo to San Antonio, and nearby points. From the very beginning, they assigned me to duty as a bugler, as they knew I had served in this role in the U. S. Army.

In the spring of 1863 we marched to Brownsville, and upon our arrival there, the battalion was merged into a regiment commanded by Col. Duff and Lt. Col. James Sweet. That year we guarded the border from the mouth of the Rio Grande in the Gulf of Mexico to Eagle Pass. Several times my company was assigned the duty of guarding the place where the Rio Grande enters the Gulf.

On the Texas side of the Rio Grande, there was a detachment of Confederate soldiers, and on the Mexican side of the river, there was a little town called Bagdad, where the customs house for the port of Matamoros was located. The war ships from the Union of the North would come as far as the sand bars at the entrance to the harbor, about two miles from the little town. Every day, the Union sailors would land in little boats that unloaded them on the Mexican side. There were also several steamboats that would arrive at Bagdad and anchor on the Mexican side. Some Union officers would take pleasure trips on these steamboats. There were also deserters from Texas and some Union sympathizers who had crossed over to the Mexican side of the river. These kind of folks would provoke us daily, shouting insolent remarks across the river.

On the Texas side of the river, there was a captain named Bruninton who belonged to our regiment and who had lived on the Rio Grande border for many years. He'd been captain of the Texas Rangers and knew the Mexicans well. He was in charge of a small patrol guarding the border on the Texas side. One day he came to our camp with ten men and delivered a Mexican prisoner to the guards, warning them to take twice as much caution with him or he would escape.

At night, Union Navy war frigates would aim and fire right at our lights. We had a two-story house, and every night our soldiers would dance with each other on the top floor, in this way drawing fire so high that the bombs would pass us by and land about a mile beyond the house. But on the night the prisoner was delivered to us, they fired several bombs that came pretty close to hitting the place. Very quickly, we extinguished all the lights. We had the prisoner's feet tied together with a chain. The following evening, the captain ordered us to move the guard to another location, because our current position was in danger of being bombed.

The prisoner was sitting by a row of corn sacks, and the guard asked him to stand up to be relocated. But the prisoner did not want to get up. As the weather was cold, he was sitting all wrapped up in his blanket. Thinking he didn't understand the order, the corporal told me to tell him to stand up, but the prisoner told me, "No, the captain ordered me to stay here."

The corporal, getting impatient, just grabbed him by the arm and lifted him up. As he stood up, we discovered that he'd taken the chain off his feet and was just waiting for nighttime in order to escape. Upon standing, he looked down at his feet and said, "Look, it came loose." This made everybody in the room die laughing, and shout and throw their hats up in the air. Captain Brunington came running at the sound of the shouts and asked the prisoner, "What's going on, man?" And the prisoner, with a feigned innocence replied, "The chain came undone."

Then the captain said to the soldiers, "This Mexican is going to get away from you, and the one who lets him get away is going to be punished severely." And some answered, "I'd like to see him get away from me!"

CHAPTER 23

That same night at about midnight, Col. Chilton came from Brownsville with a small group of men. They immediately woke us up and told us to go across the Rio Grande to capture certain men there who'd been harassing us daily. Silently, we prepared ourselves to cross the river in three small boats.

After we'd lined up ready to cross to the other side, Col. Chilton whispered our instructions to us in a very quiet voice. He told us that it was not our intention to spill the blood of any Mexican National, nor the blood of any other person if we could possibly avoid doing so, and that our only purpose would be to capture the men in question and bring them back as prisoners. There were a number of people there who were our friends, and they had furnished us the boats, the details, and the information on the men we were looking for.

Well, we had no sooner crossed the river than we succeeded in surrounding the customs house. Col. Chilton and two other officers went in and pulled out Judge Davis, who was a district judge in Texas, and a great lawyer, and after the war, became governor of the state of Texas. They also brought out a man named Montgomery who, according to what people said, was an evildoer.

Then, the soldiers found out that several Yankees were hiding out in some sand dunes close to Bagdad, and we went to capture them. About a quarter mile from Bagdad, we were going up a small hill when one of our men accidentally discharged his gun. The day was dawning and at the sound of the shot, we saw men pop up from different directions. As it was now daylight and we were on Mexican soil, we were ordered back.

In order to do that, we had to pass through the village, which by this time had been totally alarmed. So as we approached the houses, we were greeted by a rain of bullets from the houses, from the windows and from the doors. But we had received orders not to fire. Before we could reach the Rio Grande, the local judge came out to ask us why we had crossed over to Mexico. We told him we were supposed to take certain Americans prisoner, but that we had strict orders not to violate any Mexican laws.

One of our men by the name of Manuel Menchaca,[74] a native of San Antonio, mounted a horse he'd found staked out and pursued two of the men who were asleep in their hide-out in the sand dunes, where he captured them. He had a single shot pistol, and each of them had a six-shooter.

CHAPTER 24

When Menchaca got back with the two prisoners, he said to the captain, "'What do you think about Menchaquita? Will he at least be made a corporal?" Another one of us, a German who had also cut off from the group, came back with two more prisoners. Since the rest of us were already on Texas soil, and they saw the German alone on the Mexican side, a mob of men and women came up on him shouting, "Kill him! Kill him!" and someone shot him in the back with a rifle. The bullet went through him and they would have killed him had our soldiers not lined up along the river's edge and protected him. Our men fired a volley, and it knocked down their leader, who was on horseback in front, and they succeeded in dismounting him. And the rest of them fled when they saw us finish crossing over.

I came inside, cursing heavily against those who'd shot the German, and went into the room where Judge Davis was, without realizing he was there. Davis gave me such a firm look that he made me feel ashamed of what I had said, for the judge was a man of great respect.

Right after noon on the same day, orders were given to take the prisoners to Brownsville. Among these was the Mexican taken prisoner by Brunington. When everything was ready, they put Montgomery on a horse and tied his hands behind him and his feet under the horse. And Montgomery said, "All I ask is that I be treated as a prisoner of war." To which Col. Chilton replied, "You will be treated in the manner you deserve." But to Judge Davis, Col. Chilton said, "You, sir, may ride without being tied," to which Davis responded, "Thank you." Later, I found out that they hanged Montgomery on the way.

But the Mexican we had chained earlier escaped in a tragically sad manner. He and the other prisoners were going to be taken by wagon. When they were ready to leave, the guard took the chain off his ankles, and he asked to be allowed to go to the river bank, where a number of old boats were serving as latrines. The guard stood about ten feet away from him while the prisoner, all wrapped in his blanket, sat down on one of the old boats. All of a sudden, he slipped off his pants, flung away his blanket, ran to the river, and jumped in.

The guard was looking off in another direction when he heard the shouts from the soldiers who first saw it, saying "Your prisoner's getting away!" It took him so much by surprise that instead of shooting at him with the rifle he had in his hand, he threw it down, and ran after him, pulling out his pistol. As soon as the prisoner hit the water, he went under, swam underwater for about twenty-five feet, came up for air, and continued swimming under water. The soldiers who had guns at hand fired, but no one hit him. He submerged about four times. On the opposite shore of the river there were several steamboats with families aboard, and the American ladies on board began screaming. The firing increased. The prisoner was already close to the Mexican shore. After we stopped firing, they lowered a boat and picked him up. There were many Americans from up North on board and they began making fun of us, shouting harassments, and calling us cowards. They kept this up until the captain ordered the soldiers to form ranks on the banks of the river. Then he said to them, "If you start shouting again, we'll open fire." We took the remaining prisoners to Brownsville.

Meanwhile, on the Mexican side of the border, a report was telegraphed to Matamoros about our having taken the prisoners over to Texas. And two companies of soldiers reached Bagdad before the sun set that day. General Bee, who was in command of the Confederate troops at Brownsville, dispatched Judge Davis with an escort to San Antonio as soon as the troops arrived in Brownsville.

But Judge Davis' wife, who had been with him at Bagdad on the night he was apprehended, took a coach and went to ask for help from the Matamoros authorities. Almost immediately, the Mexican military authorities at Matamoros demanded that General Bee surrender Judge Davis to them, because he had been taken prisoner on Mexican soil. At that time, Mexico was at war with France, and there was a great concentration of troops in Matamoros, so General Bee had to accede to the demands of the Mexican Chief of Staff, whose name as I recall was Cortinas.[75] The following day, Judge Davis was turned over to the Mexican authorities.

CHAPTER 25

At that point, General Bee felt it would be best to send to San Antonio some of the Yankee prisoners taken the same night Judge Davis had been captured. I ended up one of those chosen to be part of the escort. We arrived at San Antonio and after delivering the prisoners, I got permission to visit my family, which was at my ranch in Bandera County.

When we returned, we found the regiment already at Santa Gertrudis, a place better known as King's Ranch. The Yankee army had disembarked at Bagdad, and the Confederate Army had retreated from the Rio Grande and was marching toward Corpus Christi.

The day we arrived back to the regiment, I was surrounded by my *mexicano* friends who were still left in the regiment (as many had already deserted). They begged me to word a petition to Col. Duff asking that we be transferred to the regiment of Col. Santos Benavides, who was in charge of a regiment of *mexicanos* in the service of the Confederate States.[76] Because during my absence, Corporal Juan Mercado had disappeared from one day to the next, and it was believed that he had been hanged.

We had in our regiment a group of about ten men who had the bad reputation of hanging folks. And it was said that they had hanged several men who didn't go along with their ideas. Among these hangmen, there was a captain named Taylor who they said had a grudge against Corporal Mercado and had threatened to hang him. He made this threat in the presence of one of Corporal Mercado's friends, a man named Juan Santa Ana. Santa Ana told Corporal Mercado about the threat. Since

Mercado had disappeared so suddenly, the boys believed he had been hanged. It was for this reason that the *mexicanos* wanted to be transferred to Benavides' regiment. They feared for their lives in our own regiment, for at every step there were constant and ugly disagreements between the *mexicanos* and the *americanos*. The colonel, however, denied our petition, telling us that we were mistaken. He assured us that Corporal Mercado had not been hanged.

About this time, Col. Duff received orders to take his regiment to the state of Louisiana. After marching about ten days, while camped on the Lavaca River, a dispute arose between a *mexicano* and an *americano*. The contention grew to the point where the *americanos* took up arms and said they would put an end to all the "greasers" in their midst. Then I and a German friend of mine, Fred Metzger, who had previously served five years in the U. S. Cavalry with me, put ourselves in between the contending parties and succeeded in calming them. This friend's name is Fred Metzger, and he lives today in Hondo, in Medina County. I believe that if it were not for him, many would have died that day.

That day, all my *mexicano* friends decided to desert that very night. They insisted that I go along with them, saying that we could take the best horses in the regiment and leave that night. I told them that if they were going to steal other people's horses, they could count me out, because if we were followed and caught, they'd hang us as thieves for sure. I told them that if I was going to go in on this, and if I got caught, I'd rather die as a deserter than die as a thief. So it was finally agreed we'd just take our own horses.

CHAPTER 26

Around that time, news had arrived that Vicksburg had fallen. Vicksburg was the strongest Confederate post on the Mississippi, and I realized that the last hope for a Confederate victory was gone.[77] That night, March 1, 1864, I, Benigno Arocha, Francisco Martínez, José Garza, José Casillas, and Santos Blanco deserted. Our plans were to go to our homes, visit our families, set a day and a place to meet again, and to cross the Rio Grande into Mexico.

The location decided on was the hills of San Gerónimo, about two miles from the ranch of Don Manuel Hernández, a man who was well-known to us all, and a friend to us all. The date of our departure was set for March 20th. On the predetermined day, Francisco Martínez, José Casillas, and I met as agreed. The others didn't show up. We waited several days for them. On the sixth day, Santos Blanco and José Garza arrived. Every night, we would go to Don Manuel Hernández' house to see if there was any news. He would help us with food, provisions, and the latest news. All day, we would stay in the underbrush on top of the hill, where there were many cedar trees and where we could keep an eye on all travelers who passed on the road from Bandera to San Antonio.

One day, I dared venture close enough to the road, to see if I recognized someone, and I saw two men go by with four camels. They were coming from Camp Verde, where the government had about fifty camels. These were going to San Antonio to pick up cargo. The men turned out to be old friends of mine. One was Henry Ramsey, the person in charge of the camels, and the other was Candelario Treviño, who'd been a co-worker of mine when I'd worked with the camels. I got within ten yards of them, still hidden in the chaparral brush, but I did not call out to them, for I felt myself to be a criminal for having deserted, and I feared being discovered.

Mr. Manuel Hernández had a big cow horn that he would blow every night to let us know the coast was clear. It was understood that if he didn't sound the horn, we would know there were people around that were not to be trusted. None of the neighboring ranchers thought it suspicious that Hernández would blow his horn because that was the way all of the ranchers called their cows. In this manner, we managed to hide out till the 29th day of March. On that day, I, José Casillas, José Garza, and Francisco Martínez set out for Mexico. Not one of us had ever been on Mexican soil.[78] We knew that we had to cross the Rio Grande in order to get there, but we did not know where it was.

We struck out in a westerly direction. We stayed away from traveled roads in order not to run into anybody. We knew that parties of Confederate soldiers were out on the road daily, looking for Indian raiders who were causing a lot of damage on the frontier[79] in those days, as well as searching for deserters who were fleeing across the border to Mexico every day.

As we weren't familiar with the terrain, we had to guess the direction that seemed closest to what we thought would get us there. After three days of travel, we were totally surprised to find a road full of the tracks of horses and wagons. We figured this was the road from San Antonio to Eagle Pass. We stopped a moment to think things over, and decided to travel on the road for a short while. We were on a high spot, and the road descended into a valley below.

We had traveled about 500 yards when we came around a turn in the road, into the bottom of the valley, and found ourselves suddenly in the middle of a Confederate company that had just stopped there. Some of the men were unsaddling their horses while others were building fires or tying their mounts. Without stopping, my companions asked me, "What do we do?" I answered, "Follow me."

The company had camped on both sides of the road, and we passed right through the middle of their camp. Some of them noticed us, and others didn't. About two hundred yards down the road, I said to my companions, "These people might try to follow us. Let's leave the road, heading to our left and separate off from each other by about 20 yards distance, to make the trail harder to follow." That's what we did, and about a half mile after we left the soldiers, we came to where four soldiers were watering their horses at a shallow pond. They had dismounted right next to a pool of water, and their reins were in their hands. They studied us closely, and since we were separated off, they stooped to peer at us between the bushes, perhaps trying to determine how many of us there were.

About a mile after passing them, we got together to discuss what to do next. It must have already been about 4 o'clock in the afternoon. Directly in front of us was a high hill with a pointed peak. Martínez was riding the best horse. It was decided that he should ride to the summit of the hill to see if there were any indications that we were being followed. The others were to travel due west, leaving the hill about a mile to the right. We could see clearly when our friend made it to the top of the hill. He descended the north side of the hill, turned around, and doubled back to us quickly. He told us that he had seen some soldiers who appeared to be following our trail, but that when they saw him at the top of the hill, they started in his direction. So this maneuver helped us gain ground in the direction we were headed. We were hoping that dusk would keep the Confederates from being able to follow our trail in the dark.

My horse was getting very tired. We kept going till midnight, and then my horse would not go another step. We had now gotten to open country with very few trees. I proposed to my friends that they leave me and my horse, and go on. I figured that if the soldiers caught up with me the next day, I could hide somewhere in the brush. And if they found my horse, they would surmise he'd been left behind when he tired out. And maybe by the following night, I could travel with my horse until I got him to water. Some of them agreed to this plan. But Francisco Martínez said no, it was better for all of us to die together if our time had come, without abandoning one another. So we stayed there that night.

CHAPTER 27

At dawn we began to see herds of wild horses. One of these herds was about a mile from where we stood, and one of the horses was by himself about fifty yards behind the others. Martínez said that one was a tame horse and, mounting his horse, he rode off to see if he could rope him. Since we were on a flat plain, we could see when he reached him, and his horse stopped. Then Martínez dismounted, put the rope over him, and brought him back to where we were. He was an old government horse that soldiers on some expedition against the Indians had abandoned because he'd tired out. He was very thin, but he wasn't tired. I put my saddle on him, and we continued our journey.

We no longer had any food left. About two miles from where we'd slept, we came upon a young calf maybe just a month old. The poor little animal was asleep in the thick grass. Our companion Francisco Martínez got off his horse, knife in hand, and crawled along very carefully so the calf would not hear him. When he got close enough, he plunged the knife in the calf's heart. I couldn't help but imagine Martínez as a lion and the little calf as the innocent victim.

Each one of us took as much meat as he wanted, and we kept going, hoping to stop to eat as soon as we could find water for ourselves and for our horses. About eleven o'clock in the morning, we got to a deep creek bed with a few standing pools of water. We decided to camp down in the creek bed and keep our horses there, so they would not be seen should someone happen along. There was a little grass along the creek bed and we figured that was enough for the horses.

Among the many grandchildren who carried on the Tafolla *vaquero* tradition are the two shown here—Manuel, standing, and Jimmy, third from left, on horseback. (Courtesy of C. Tafolla.)

We built a fire and started roasting meat. After eating, I saw that the horses had tread on what little grass there was and they didn't want to eat it anymore. Seeing how much grass was outside the creek bed, I suggested we take them out of the creek canyon so they could eat their fill. Each of us could stand guard an hour, and take turns relieving each other. Jose Casillas' turn was first, and I was to relieve him.

I walked about 200 yards west of the horses and found a well-traveled road running north and south. It had wagon tracks and tracks of troop horses, and I figured it was the road from Eagle Pass to Ft. Clark, because both of these forts had troops. Immediately, I notified my companions that we were in danger of having a troop of soldiers pass nearby. We decided to hide the horses in the hollow of the creek again and wait until later in the day to get going.

It must have been about four o'clock when I peeked outside the banks of the creek bed and saw about ten or twelve men in Confederate uniforms coming straight toward us, apparently following our trail. They were about 300 yards away.

We grabbed our weapons and prepared to fight. Since they couldn't see us, we had a clear advantage. We had no time to waste. We thought we should let them get very close, and then fire together, each of us singling out one and firing, and continuing to fight until we either died or won. We were in a position where we could see them clearly, and they couldn't see us until they came right up to us. They were about fifty yards away when the man in front, who appeared to be the leader, stopped and spoke to the others. Then they changed course and went straight north. We kept watching them till they were lost from sight. This seemed to have been an act of Providence.

CHAPTER 28

It was about five o'clock in the afternoon when we left this place and headed west. About twelve o'clock at night, we started to go downhill on a very long slope, and I told the men I thought we were getting close to the Rio Grande. Soon after, we came to a bamboo thicket down a deep and narrow path. I walked in front, and my tired horse behind me, and the men followed. We walked along this path through the bamboo reeds and the wooded underbrush without being able to see anything but the stars. At last we came to the riverbank, and I stopped. And I said to my companions, "This is the Rio Grande. But now, how do we get across?"

Some of them doubted that it was the Rio Grande, but I insisted, "This is it. And as soon as I get to the other side, I'm going to shout "*¡Qué viva México!*"

We got to the very edge of the water, and I said, "Wait for me, I'm going to find out how deep it is." I took off all my clothes, and picked up a stick to use as a staff. It was February, and the water was very cold. The current was very swift and almost swept me away, but I fought it till I got to an island in the middle of the river. I crossed the island, and got to the river branch on the other side. But it occurred to me that my companions might not be safe, considering that a squad of soldiers might come by since they patrolled the river day and night. I went back and helped them cross to the island, and then I went to check the depth on the other side. It was just as deep and just as strong a current, but when I got to the other side, it was at a point with a very high bluff, and because it was dark, you couldn't see where the bluff ended. I followed the edge of the water for about a hundred yards and finally found a very difficult path up the bank of the river, but it seemed to be the only way up. So I went back and brought my companions. We came up the path on foot, leading our horses by the bridle reins.

As soon as the last man had crossed, I shouted "*¡Qué viva México!*"[80] Dogs began to bark in response. And not far away, we laid down to sleep as if we were at home.

At daybreak, we discovered that we were in a field of crops, and that there were some houses about 400 yards away. I asked my companions to wait while I went to look for something to eat. People were just beginning to stir at the home I came to. I greeted them and told them we were Confederate deserters and wanted to buy something to eat. The man of the house asked if any one in our group was from Bandera County. I answered that I lived there. He asked me if I knew Don Andrés Gonzales. I answered, "I know him very well, both he and his family." And I gave him a report on each one of them by name, for I knew them all well. This gentleman received us very affectionately.

After breakfast, we went to introduce ourselves to the mayor of Old Monclova. This gentleman turned out to be a relative of my wife. He received us very courteously and gave us a passport so that we could cross to Piedras Negras. He advised us to leave our weapons with him, because since Mexico was at war with the French, no foreigners were allowed to carry weapons. Later we returned to pick up our guns.

After being in Piedras Negras for a while, I sent for my family, and we went to live at Presidio del Rio Grande, which is now known as Guerrero de Coahuila.[81] There I found the family of Corporal Juan Mercado, the man we thought had been hanged by Captain Taylor's gang. We found out that he had deserted to Mexico, and had sent for his family which was then living on the Medina River close to San Antonio.[82] After being with his family for a month, he committed suicide. They say that he went crazy, and in a state of depression, took his own life.[83]

I lived for a year in this village of the Rio Grande, and during this year, I became acquainted with an *americano* doctor named Cooley.[84] This doctor invited me to go along with him as an interpreter as he visited towns in the state of Coahuila. He promised to feed me and my horse and to pay me a dollar a day. We went through several villages but found no sick people. One day he said, "If I could find a rich Mexican with an intermittent fever, I'd get all our expenses back from him. I'd give him a dose of medicine which would make him feel like he was dying, and I'd keep him in this condition for a few days until I'd get enough money for all our expenses."

The following day, we arrived at the village of Nava and there lived a rich man named Don Felipe de la Garza. His son-in-law told us that his father-in-law was sick and would like for the doctor to come by and see him. He took us there himself. Don Felipe de la Garza was an elderly gentleman, very fair-skinned and very fat. You couldn't tell he was sick. He was seated on a very high, very richly decorated bed. His wife was on another bed, caddy corner to his, in the same room. She also was very fair and fat. Both appeared to be the same age, about 65 or 70.

Mr. Fulgencio de la Garza, the son-in-law, explained Don Felipe's illness to the doctor. It was in his legs. For fifteen years, he had been unable to walk. Several doctors had seen him, but none had been able to cure him.

Dr. Cooley examined him and said he thought he could cure him. If he would like to put himself under his care, Dr. Cooley proposed to compound him a medicine which would last him three weeks or a month, after which time he would come to see him again. Don Felipe agreed to the doctor's offer. But his wife, who was just listening to the conversation, told her husband in Spanish, "Ask him how much he is going to charge you. Watch out that he doesn't poke one eye in, and take the other one out."

The doctor answered that he would furnish enough medicine to last him for a month for five dollars. Upon hearing this, the lady said to her husband, "Didn't I tell you? Tell him no, tell him to sell us two bits worth of the medicine, and if you get better, you'll buy more. If not, you won't."

With that, we bid them adieu. As soon as we were alone I said to Dr. Cooley, "What do you think about rich Mexicans now?" He responded that they were poor prospects.

We spent a month traveling through the villages in the state of Coahuila, but found only one lady who was sick, in the village of Zaragoza. We lodged in the home of Don Anastacio Gómez, whose wife happened to be my wife's aunt. The sick lady was Don Anastacio's daughter-in-law. We stayed at their home for four days. They took good care of us and fed our horses well, but would not accept any pay, hoping that the doctor would not charge for the sick woman's care. But the doctor charged them nine dollars. They complained that it was too high, but the doctor responded that, had it not been for the kind treatment we had received from them, he would have charged fifty dollars.

This was the only money he made during the whole trip. We went back to Guerrero, and I never saw the doctor again.

CHAPTER 29

At that time there were two *americanos* who lived on a farm they had rented three miles from town. These men had left Texas to escape the war and had planted crops on that ranch. That ranch was located on the road that goes from the Rio Grande to Nava, and since this was the route generally taken by the Mexican troops, and they spoke no Spanish, the soldiers from the Imperial Party had come through and wiped them clean of everything they had. They didn't even leave them anything to eat or anything to sleep on. About a month later, another group came through and once again took everything they owned. One of them was completely discouraged and he left. The other one came and asked me if I'd go live on the ranch with him and he would give me the other's share, which was half the crop. Nearly all the work had been done. I went. It was already June of 1865.

In August, Freeman, as the *americano* was called, who had heard the [Civil] War was over, begged me to lend him my horse so he could go see his mother, who lived in Halletsville, Texas. He told me that if he failed to return, I should take his share of the crop in return for the horse he had borrowed. This is exactly the way it turned out to be.

By the end of October I had finished harvesting the crop, and decided to return to Texas with my family. We crossed the Rio Grande at Piedras Negras (now called Ciudad Porfirio Diaz)[85] in a wagon drawn by a yoke of oxen.

After traveling for four days, we stopped to take a siesta on the Frio River close to a ranch owned by an *americano* named Griner. Soon after we got there, a young Negro came along driving a herd of horses, among them a mare, a three-year-old mule, and other horses which the Indians had stolen from my ranch two years before.

I asked the young Negro where they had bought these animals. He told me that the Indians had left these tired out animals, and had taken Mr. Griner's horses. I went to visit Griner and it turned out that we already knew each other, for he had been in the same company with me, in Col. Duff's regiment, from which I had deserted. After he recognized me, he turned over my animals to me without the least hesitation. This was really a piece of good luck on our trip, for I had never thought I would see my animals again.

I arrived at my ranch at the end of October. And since my brother-in-law and his family[86] were living there, I left my family in their care while I went to Camp Verde, where the government camels were, to look for a job. At that time, they were under the charge of my good friend, Fred Metzger, and, since he needed an employee, he gave me a job right away. I stayed on this job till June of 1866. At that time, the government sold the camels, and I contracted with a widowed lady who owned a herd of cows and offered to give me one half of the calves born in return for my labor.

In October 1867, I made a deal with my friend Metzger to go live on a ranch he had bought on North Creek close to Comfort and close to the Guadalupe River, in Kerr County. That year I stocked the ranch with about a thousand head of goats.

My mother-in-law was living with us at that time. In the month of November, she became very ill, and died. My brother-in-law, Blas Torres, had come to see her and he, my wife, and I were at her bedside when she passed away at dawn on November the 25th. On that day while my brother-in-law made the coffin, I drove to Comfort, about five miles from there, to make arrangements for a two-horse carriage. And we left the ranch with the body, on our way to Privilege Creek, twenty miles away, about 7 p.m.

The ranch was between some very high hills, and you could see the moon over the tops of the hills, so bright and so beautiful that I'll never forget that night whenever I see a full moon. My wife and I and our six-year-old daughter rode in the same carriage. My brother-in-law rode in front on his horse. In those days Indians still criss-crossed that part of the frontier, stealing horses and killing everyone they found. And they would generally make these raids when the moon was full.

That was one of the saddest nights I've ever experienced in my life. We had to cross the Guadalupe River and travel over a rough and lonely road. Upon leaving the ranch with the body, we had commended ourselves to God and I know He was with us, blessed be His Holy Name! We arrived at the ranch in Bandera County where my brother-in-law lived at about five a.m. The sun was coming up already. His wife was alone waiting for him. And when she heard the sound of the carriage, she got up and ran outside, asking for my mother-in-law, because she knew she'd been ill. She looked at the carriage, and as soon as she saw my wife and the coffin, she understood what was happening, and hugged my wife, weeping. After burying my mother-in-law, my wife and I returned to the ranch on North Creek.

In the month of February the time arrived for my wife to be confined, as she was almost ready to give birth to a child. I begged my brother-in-law to let his wife come to stay with mine during the period of her illness. But he refused, so I had to take my wife to stay in his home which was a *jacal* so badly constructed that it was impossible to keep the cold air from penetrating the walls. At that point, a season of snow set in, and a very cold "Norther" that lasted about fifteen days. Since the house offered such poor protection from the weather, my wife got pneumonia and died.[87] The doctor couldn't save her. An *americana* assisted the doctor in tending to my wife. Had my wife been at my house, I believe she would not have died, as we had a very good house. I was left with three girls and two boys.[88]

This was in February of 1868. That year I came to live on the ranch of Don Miguel Indo, on the San Antonio River, twenty-five miles downstream from the city. I planted crops there that year, and when we finished planting, I began to travel with a wagon train, [carrying freight] to the ports of Powderhorn and Indianola.[89] This was before the railroads came to San Antonio.

In the month of May [of 1869], in San Antonio, I renewed my acquaintance with Señora Anastacia, the widow of corporal Mercado. In the month of June, I married her.[90] She had two children from her former marriage, a son and a daughter.[91] After the wedding, we went to live on Don Miguel Indo's ranch, where I had my crop planted.

On the 4th of July, the San Antonio River went on a rampage, flooding all the fields and ruining all the crops. After the flood, there was an epidemic of malaria fever in all of the villages and farms which had been inundated. My family and I became ill, and my mother-in-law and my

This tin-type of Anastacia Salinas de Tafolla was taken in 1869, a few months after her marriage to Santiago June 14, 1869. (Courtesy of C. Tafolla.)

wife's other relatives came and took us with them. I was so sick I didn't even know when they removed me from that house. The whole family and I were taken, and I didn't know it until that night, when my fever went down. They took us with them to a ranch on the Medina River. That night we camped out, and the following day arrived at the Medina River, to the Herrera ranch, which is called El Paso de los Garzas.[92] There I stayed for a whole month, so sick I was unable to work at all.

About this time I heard that a Jewish merchant named Gold Frank was looking for a man to travel to Mexico with him. Mr. Frank traveled in a buggy drawn by a pair of mules, and he took two men, one to drive the buggy and another on horseback. One day, one would ride the horse and the other would drive the buggy, and the next day, we'd change places. Since I was sick with chills and fever, I suffered quite a bit on the days I rode horseback. And the days when I rode in the buggy, I rested. Gold Frank's business in Mexico was to collect money owed to him by merchants, as he would supply them with merchandise from San Antonio, and each year make a trip to receive payment. We crossed the Rio Grande at Eagle Pass, and went to Zaragoza, Morelos, Nava, Allende, Santa Rosa, and Monclova, and then returned home.

After we arrived back in San Antonio, I decided to go home to my ranch on Privilege Creek in Bandera County. This was in November of 1869. In 1870 I planted on my ranch and gathered a good crop of corn. The following year I moved to the ranch of John James, which was only one-and-a-half miles from Bandera City. I lived there two years, and in 1874 I went back to live on my ranch.

J. P. Rodríguez was the Justice of the Peace at that time for the precinct of Privilege, as we called it. That year I became a traveling merchant. I bought goods in San Antonio, and got a license to do business in Bandera and Kerr Counties. I sold all kinds of merchandise, groceries, and clothing. I would also buy cowhides, and hides of all kinds. I'd also sell liquor, but I did it without a license. I had certain customers who ordered it in advance and would sell me hides at very low cost, because they were stolen.

Family Photo, circa 1887, Santiago, Anastacia, and children. All of the clothing in the picture was sewn by Anastacia by hand. Missing in this picture are Santiago's five oldest children (by Juanita Torres de Tafolla): Pete, Molly, Juanita, Catarina, and Santana. Standing (left to right): Ernesto, Santiago, Jr., Mariano F., Anastacia's half-sister, Gabriel, J.J. Mercado, Jr., María Mercado. Seated (left to right) Santana, Santiago Sr., Anita, Anastacia, and Lott. Standing to their right, J.J. Mercado III, son of Anastacia's son J.J. Mercado.

CHAPTER 30

In those days there were lots of herds of unbranded livestock running wild in the frontier, and I would make trips even farther up than the headwaters of the Medina and the Guadalupe Rivers, where there were no permanent settlers. There were, however, groups of *americanos* who camped for weeks and months at a time, hunting deer, bear, and wild turkeys, and searching for beehives full of honey. But they would kill more cows and yearlings, bulls and calves, for the purpose of skinning them and selling the hides. I would visit these camps accompanied by one of my wife's relatives, whose name was Jacinto Treviño.[93]

On one such occasion, we came upon four *americanos* with a cart drawn by a yoke of oxen and loaded with hides both fresh and dry, and fresh and dry meat as well. They were coming down a very rough and rocky trail. There was no road, only the trail of their own cart, which we were following.

As soon as they saw us, they grabbed their guns and asked what business brought us there. I answered that I was selling goods and buying hides, that I had been told there was a camp of hunters and so we were looking for them in order to sell them something and buy the hides, if they were interested in selling them. They answered that all their hides were stolen, because they killed cattle of any brand. They figured that the owners of these animals would never look for them in these desolate areas, because the owners lived where the land was thick with settlers and without danger, "while we live here on the frontier, exposing our lives, as well as those of our families, to the risk of being murdered by the Indians."

This lengthy harangue was delivered in a threatening tone, and with their hands on their weapons, and I was careful not to contradict them in any way. I answered them that I was not interested in anything except selling the merchandise we had, and buying hides. But they said that they didn't trust us. Again I assured them they were not running any risk, that I wanted them to be my customers and I wanted to be their friend.

To prove this point, I pulled out a bottle of whiskey and offered them a drink. That picked up their spirits, and they invited us to their homes. They lived near the headwaters of the Medina River. As soon as we got there, they bought a gallon of whiskey. And they started drinking and buying our merchandise in exchange for hides. We camped about a hundred yards from their houses. They brought their families there to us, and bought chintz fabrics, heavier fabrics,[94] and all kinds of items which we stocked. They sold us about a hundred dollars[95] of hides, and beehive honey, and beef fat. And they bought more whiskey, and ordered many things for us to bring on our next visit.

As soon as it got dark, they began to play the fiddle and dance and shout. About ten or eleven o'clock, one of the oldest among them came to our camp and told us, "Friends, it's better that you get out of here right away, because the boys are getting drunk and they want to come out to kill you." At that hour of the night we hitched our horses to the wagon and left. We never saw them again because we were afraid to return.

On another occasion we went looking for hides on some ranches by the Medina River, from Bandera on south. We got to a ranch house that stood alone on top of a tall hill. The doors of the house were all closed, but it was evident that people lived there. Signs of fresh wagon tracks led from the place to some distant, cedar-covered hills perhaps three or four miles away. My friend spent quite a while looking around the grounds surrounding the house, and about fifty yards from the house, he found a pile of raw hides in the chaparral. He came to give me this information and we decided to follow the wagon tracks in order to meet the people and try to buy their hides. In keeping with this plan, we got on our wagon and followed the trail about four miles through the cedar-covered hills. It was cold and we wanted to build a fire when we saw smoke about a mile ahead, in the direction we were going. We decided to keep going till we got there.

We had traveled about a hundred yards when we saw a man, a woman, and some children. I told my friend, "Stay here. Let me go find out what they're doing." When the man saw me coming, he grabbed his rifle, and came to meet me. About fifty yards away he stood and asked, "What are you looking for here?" I answered that I thought I was looking for him, since I'd been at his ranch, had followed the wagon tracks, and was trying to find him to sell him some merchandise. He asked what kind of merchandise we had, and I invited him to come see for himself. He went with me to the wagon, and the first thing I did was to offer him a swig of whiskey. He drank it, and then invited us to come to his camp to warm ourselves [by the fire].

When we got there, the lady seemed to have misgivings about us, but she didn't say anything. The man said, "This is the way I make my living. There's a lot of livestock in these hills that belong to people who live far away from here, in well-settled, civilized regions. They'll never see their animals again, and here we are, open to getting ourselves killed by Indians. I can't leave my family alone in the house. I take them with me when I go to kill a few cows or yearlings, to sell the hides and help me make a living."

After a while, we became friends, and I tried to buy his hides. And he said, "I only have these here." He had about six dead yearlings, in a trench about three hundred yards long, which he had shot that morning, and he was working at skinning them. But by the time we got ready to leave, we had become good friends and I'd gained his confidence. And he sold us not only the six yearling hides, but the whole stock of hides he had hidden close to the house. From then on he became one of my best customers.

During the time that I was dealing in hides and selling merchandise, liquor became such a vice to me that I was controlled by the liquor habit. And, at the same time, I began to lose my sense of respect for what belonged to others. I would buy stolen hides from all my neighbors on Privilege Creek regardless of whether the animals belonged to them or not. And I would treat everyone the same, whether *americano* or *mexicano*, and say to them: "Bring me all the hides you can get, regardless of what brand they have, even from my own cows." This encouraged them to go after animals that didn't belong to them. Some of them would bring the hides to my home at night and sell them to me at very low prices. A hide worth three or four dollars on the market would be exchanged for a bottle of wine and fifty cents worth of coffee. I had gained the friendship of all the wicked men in the neighborhood and was well loved by all.

When I had collected as many as fifty or a hundred hides, I would take them to San Antonio to sell them. They had to pass inspection before leaving Bandera County, but I had the inspector paid off. After paying him ten cents on each hide, I would give him a bottle of good whiskey, and I had him recording on his inspection list any brand I called off to him, even if the brands on the hides weren't actually the same [ones I'd said they were]. All that and much more can be done when controlled by liquor. The inspector was happy to say, "Live and let live," and neither he nor I would listen to the voice of our conscience. Also, in order to keep my neighbors happy, I'd hold dances in my home for them.

At that time, J. P. Rodríguez was Justice of the Peace of the electoral precinct. But he wasn't very popular, because he was trying to stop the sale and traffic in stolen hides. And although he was always with us at our dances, horse races, and other lawlessness, we had to watch out for him, so he wouldn't find out about our killing animals belonging to others or trading stolen hides.

In 1875 when election time came around, my friends nominated me as Justice of the Peace for the precinct. And, as I was wicked, they wanted me to be the judge. On election day, Mr. J. P. Rodríguez did not get a single vote.

In due time, after the election, they called me to Bandera to take the oath of Justice of the Peace for the precinct.[96] And when they asked me "Will you swear to obey and see that the laws of the state of Texas are obeyed?"

Editor's Note: Although the manuscript ends here, a clearly controlled pencil mark on the next line of the manuscript suggests that the author began to write more, but perhaps hesitated to consider what would come next.

EPILOGUE

Santiago Tafolla ends his autobiography abruptly in 1876. We are left in suspense, not knowing several important facts about the last thirty-five years of his life. We as editors, having examined the memoir's original manuscript and noting his detailed style of writing, do believe it was Santiago's intention to write more about the huge transformation in his life that occurred in 1876.

This epilogue will attempt to describe the events of Santiago's life from the Justice of the Peace swearing-in ceremony in 1876 to October of 1911, when he died at the pulpit in Seguin, Texas. Given the high demand for him in his chosen field of work, Santiago, unfortunately for us, did not return to the conclusion of this memoir. We do, however, have many important and well-documented facts from which to draw our conclusions. Some of those documents came from his own hand, some from the writings of his friends and companions, like Policarpio Rodríguez, and some from church records.

The records of Bandera County attest to the fact that he did serve as Justice of the Peace of that county, having posted bond for the position on May 29, 1876, but they also record the official acceptance by the court of his resignation from said office within months, with the formal acceptance of his resignation occurring on December 4, 1876.

> It appearing to the Court that James Tafolla had resigned the office of Justice of the Peace for precinct # 2 of Bandera County, and his said resignation having been accepted, it is ordered that

Camp meeting c. 1901 with Santiago Tafolla, seated. His youngest daughter Anita is standing to his right with her hand on his shoulder.

Santiago and Policarpio at the Conferencia Fronteriza Mexicana of the Methodist Church, San Antonio, 1900. Santiago and Policarpio are in the middle row, standing and white-bearded, Policarpio at the center and Santiago second from the right.

T. F. Randolph be, and he is thereby appointed, Justice of the Peace for said precinct.

We also know that soon after he was elected to the office, he joined the English-speaking congregation of the Methodist Church and began teaching and preaching in Spanish to the Mexicans in the Privilege Creek community. A. H. Sutherland had been commissioned by the Methodist Episcopal Church South to promote the work of the Church along the Mexican border on both sides of the Rio Grande. He learned from reliable sources about a promising "young man" who was instructing and preaching the Gospels to the Mexican people in the Privilege Creek community of Bandera County. He dispatched José María Casanova, one of his young preachers in San Antonio, to visit Tafolla. This was done more than once and Trinidad Armendáriz, another young and talented preacher, soon included the Privilege Creek community in his itinerary as well. The "American congregation" was very much impressed by the courage and enthusiasm of Tafolla, and recommended he be given a license to preach. In October of 1876, the Annual Conference which met at Seguin, granted him a pastoral license and assigned him on trial as pastor of the Mexican congregation at Laredo, Texas. This must have been the point at which Santiago asked to be relieved of his duties as Justice of the Peace in order to undertake his first pastoral assignment. This assignment was so important to Santiago that he mortgaged his entire ranch in Bandera County in order to fund his family's move to Laredo, for the Church provided no provisions or funds to sponsor his relocation, and he was already supporting a family of seven, including three young children and two stepchildren. (He was also the father of five only-slightly older children by his first wife, Juanita, who were being cared for by Juanita's relatives.) He put his family in a covered wagon drawn by two horses and undertook the long and difficult journey to Laredo, where he arrived in March 1877. This was before the advent of the railroad to the border city. It was also the first official appointment made by the Southwest Texas Conference of the Methodist Episcopal Church South to the Mexican border.

This is what the official records report; however, there is an emotional and difficult transformation which is not recorded in the official documents. We know from his memoir that Santiago had long prided himself on his honesty, had quit an apprenticeship for this very reason, and had often found distaste in the actions of others, even his supervising officers, when they did not act with parity and justice. Yet he finds himself swearing (on a Bible, most likely) that he will do the very thing which his "friends" have elected him not to do. At the very moment he is asked to swear that he will uphold the laws of the state of Texas, he is filled with—remorse? shame? surprise? disgust? a sense of lost integrity? We do not know the specifics, but we do know the transformation was total and happened over a very short period of time. By the time A. H. Sutherland sent someone to investigate this "young man," Santiago had already gained a reputation for preaching within his community. In addition, we know of his attempts to convert Policarpio and the other members of his community organization.

Policarpio, in his own memoir (1898), tells us that Santiago was the president of an organization which was "for mutual help and instruction." With a constitution and by-laws, an opening prayer, and Policarpio as secretary, they had debates and discussions and Policarpio admits saying to Santiago, "Something great is going to come out of this society; no telling where it is going to" (89). It must have been very close to the time of the swearing-in that the Methodist preacher José María Casanova came to the area, and even showed up at the meeting of their society. Policarpio writes, "We accused Tafolla of bringing the preacher there, and the disagreement among us broke up the society. Soon after this, Tafolla joined the Methodist church" (90).

Santiago's renewed commitment to his faith causes problems in the relationship with his friend and cousin Policarpio. Suspecting that Tafolla had become a Protestant, Poli asks him, and Santiago admits it. In response, Poli tells Santiago he is no longer welcome at his house. But Santiago is too intent about his mission and his friendship to stop there; he finds the perfect bait for Policarpio's interest—a horse—and when he next encounters Policarpio on a road, and Poli tries to pass by without speaking, Santiago calls after him that he has a filly for sale at a good price because he will soon be leaving (probably to Laredo, which places this event between October 1876 and February 1877—most likely toward the latter end of that period.) Policarpio agrees to come by the next day, and Santiago is ready, with a preacher named Trinidad Armendáriz about to conduct prayers at his house. Angry at having to listen to them pray for him, Policarpio gets out as quickly as he can, but he spends a sleepless night and is then "wretched and miserable" (91) the next day, wanting neither to drink with his buddies nor eat at their dinner. That night, he falls down on the ground in prayer, and emerges from it a joyous, if crazy-sounding, man. He runs home and kisses his wife,

Anastacia with her two sisters, a half-sister, and a niece. Anastacia is standing on the left, her sister Sabinita standing on the right. Identity uncertain of the woman seated far left.

strips all the saints off the walls, and reads the Bible. Policarpio recounts that his wife, upon witnessing his strange behavior, "burst into tears and told the children that their father had gone crazy" (94). Within months of each other, Santiago and Policarpio were both wholeheartedly involved in Methodist evangelism among the *mexicanos* on both sides of the border. In 1877, Policarpio received his trial license to preach, and in 1882 he built at Privilege Creek what is now known as "Polly's Chapel" from limestone he had cut and hauled with his own hands.

For several years in succession, Santiago filled some of the most important pastoral charges on both sides of the Rio Grande. His "hoja de servicio" shows his pastoral duties as starting in Laredo, going for several years to Corpus Christi, then San Diego, Guadalupe, and San Antonio, Texas; Monterrey and Lampazos, Mexico; then Tobey, Pearsall, San Marcos, and Seguin, Texas, among other, unrecorded places. While still a "new" preacher, on his first "permanent" assignment in Corpus Christi, we know that his church reported in 1878 a membership of only 45, but an impressive 250 people attending services!

In 1881, he was appointed by the Annual Conference as Presiding Elder, from the San Diego Mexican Mission District. A. H. Sutherland says in his writings that this was the first time in the history of the Methodist Church that a preacher of Mexican descent had ever been elevated to this position of trust and responsibility. He had the distinct honor of serving two more times in this capacity during his ministry. Twice he was elected by his Conference as Clerical Delegate to the General Conference. At the West Texas Conference, in Gonzalez, Texas, Santiago was named President of Border Missions in Texas.

But his encounters with local communities were not always pleasant. In the primarily Catholic communities in which he preached, Protestantism was seen as worse than even the Papally condemned Masonry. A community dispute in Corpus Christi over the validity of local schools and their openness to Mexican-American students proved to be a thinly veiled "civil war" between community Catholics and community Protestants. The following interchange, between Santiago and the editor of a newspaper in Corpus Christi, Texas, early in his preaching career, shows evidence of this friction.

An article in the Spanish-language newspaper *El Horizonte* expressed doubt as to whether the public "American" schools would accept Mexican-American children, "tiempo perdido mandar a nuestros niños a las escuelas americanas." So Santiago, newly arrived to town with three school-age children, goes to see the mayor "con algún temor de ser avergonzado" and asks him if this is true.

> But the mayor assured me that not only would they receive Mexican children gladly in the public schools, but that he would personally accompany the parents to enroll the children, which is what he did with me. . . .
>
> Any honorable man and good citizen should pledge their efforts for the education of our youth. For only in this manner will the government end up having good citizens. But I see in this article complaints against the administrators of our public school system. . . .

With that letter, (and with his vocal presence in the community) he drew the ire of the editorial staff, and a long duel of newspaper letters followed between Santiago and Santos López, where Santiago writes "Usted no ha olido la doctrina de Cristo pues pertenece al sistema esparío de la iglesia romana" and Santos López responds, "¿Dónde aprendió a espresarse así, don Santiaguito? ¿Sería en alguna cantina de San Antonio?" He also accuses Santiago of preaching falsehoods, because "it is all based on absurdities, a natural fruit of the Protestant reform." The arguments descend from there to a mocking verse about a Tafolla who "confuses error with truth . . . ignorance with science . . . and vice with virtue." Much of their dispute over educational systems appears to be mere preface for the deeper argument between them—the conflict between the two religious traditions. By, in effect, "integrating" their *mexicano* religious community, Santiago upset the status quo, and challenged traditional community power structures.

By now, Santiago and Anastacia had seven children in the household, and more on the way. Dedication to preaching *el evangelio* was total. The same fearless energy which he displayed as a young soldier was now evident in his enthusiasm as a circuit-riding preacher, and would continue through his forties, fifties, sixties, and seventies. Rev. Pedro Grado tells of one occasion when they left San Diego together in bad weather, because Santiago wanted to keep his promise to preach to some folks that night. A torrential storm left a rushing creek bed rising before them, as violently turbulent and murderously dangerous as the Rev. Grado had ever seen. They hesitated—Santiago seeming to be in deep meditation or prayer—and then Santiago, no longer a young man, asked him, "Do you know how to swim?" When Grado nodded, Santiago was relieved and answered, "Well then, I won't have to go back on my word. I promised Brother X that tonight I'd preach at his house, and dead is the only way I won't do it." And with the reins to the one-horse buggy in his hands, he headed into the water. The horse pulled free, but the buggy was stuck in the rushing water. Santiago dove in, calling to Rev. Grado, "Dive in. Don't be afraid." Santiago emerged at the other side, having swum through the currents unharmed, and urged Grado on until they were both safe, at which point they both made plans to have the buggy pulled out as well. This was not the only instance of punctuality and obligation which Grado cites of Santiago's commitment to his promises.

Santiago, while filled with faith, energy, and purpose, was also a human being with all of the family problems and predicaments inherent

Rev. Ernesto Anastacio Tafolla, at his wedding with Esther Margaret Colín, c.1907

in being a husband and father. He took seriously his role as father to all fourteen children which he and Anastacia considered theirs. Upon marrying, he and Anastacia already had seven children, his five by Juanita, and her two by J. J. Mercado. They proceeded to have seven more children of their own (Gabriel, Santiago, Jr., Mariano F., Ernesto, Lott, Anita, and Santana.) While he was ever the attentive pastor and often called upon to help others in need, his own family required support as well, and economic situations were difficult. A. H. Sutherland re-affirms that, as far as material things were concerned, Santiago was poor, but he was filled with an incalculable wealth of character, energy, faith, and a characteristic independence.

In addition, Santiago suffered the death of two of his sons—both named Santana. The first Santana, son of his first wife, Juanita, was gored by a bull as a boy and died from the wounds. The second Santana, his son by Anastacia, died as an adult, a married man who had an affair with and fathered a child by a woman named Argentina. (According to Sarah Tafolla—one of Santiago's grandchildren—the child was Hazel Patritus, mother of concert pianist Martha Patritus.) Torn between his allegiances and unwilling to leave his mistress, Santana committed suicide, and Santiago, his father, found the strength to deliver the sermon at his graveside.

Still active, traveling, and in demand as a preacher into his 70s, Santiago led a busy life. We know that he kept a pocket journal from 1909 to 1910, in which he recorded his salary ($36 a month), the bills he paid, his visits to and letters from his children, which ones were answered and which not, and which cities he preached in and how many times, as he traveled to Luling and back, to La Vernia or Schertz, or more often, to the house of some *hermano*, to preach and then return. He kept track of when he left his watch off for repair and how much he paid for firewood or horseshoeing, and three times in ten pages, he mentions books: a request for a Spanish dictionary from his son J. J. Mercado, Jr., a teacher in Houston; the order for a Biblical dictionary to be sent to a fellow pastor; and on the 31st of January, a happy receipt by mail of the *Almanaque Cotidiano para 1910* from Spain. He also mentioned paying for a newspaper subscription, stamps, and ordering his clergy certificate, said application of which gets returned to him because it is missing the signature of the Presbítero Presidente, and the same day he complied and re-sent the application by certified mail.

Despite his ministerial duties, his family is an important part of his daily concerns, and much attention is focused on the letters from his children, or his own letters to his wife, who did not always accompany him on all the pastoral trips. There is a tenderness obvious in the pattern, as he writes to her within a day of her leaving to go back to San Antonio, and frequently thereafter, and there seems to be a reluctance to say goodbye, especially in one humorous entry, "[My granddaughter] Odessa and my wife left. I barely escaped killing myself as I jumped off the train." He mentions buying a turkey for "thanks given day", which he and

Marianito, the eleven-year-old grandson of Santiago, in 1908, the year Santiago began this autobiography. It was said by many family members, Santiago among them, that Marianito was the grandchild who most resembled him. Marianito, who was eleven in this photo, died three years after this photo was taken, the same year as Santiago died. The editors felt that this was the closest resemblance we could present the reader to what Santiago must have looked like in 1848, when he left New Mexico at the age of eleven.

Anastacia would spend with their son Mariano and his family, and with Anita and Santiago and their spouses. His entire thirty-five-year career as a pastor seemed to have every day been filled to the brim with professional and personal responsibilities.

But on September 16, 1908, Santiago sat down to write his memoir, and proceeded smoothly and eloquently, with one single draft of what he wrote. We know he felt unschooled in how to go about such a task, and lacking the editor, transcriber, and translator which Policarpio had, he set his own format, probably leaning heavily on Policarpio's book as a model. We even suspect that the reason he stopped writing after the mention of the swearing-in ceremony was that he had been told or had surmised that a book should only go 30 chapters, and he was wondering how to summarize all the rest that remained to be said.

What motivated Santiago to write these memoirs? Yes, his friends and children had asked him to, but Santiago had always chosen his own path and his own priorities. Certainly, he wanted to leave the knowledge of what he had seen and experienced to his descendents and friends, and, undeniably, he was a natural evangelist, with thirty-two years of experience preaching at the time he began this book. Certainly he wished to show others the clear "Providence of God" and the divine ability to transform even those lives which had been overrun with vices and "pride" like his own. The memoir itself is a confession of prideful and sometimes dishonest actions.

But perhaps there were other things as well which Santiago wished to accomplish with the telling of this story. As we read the precision and the urgency in his writing—naming names, places, and events, and giving precise dates and details—we were aware of a certain inherent insistence that this was THE TRUTH, that he was not merely an old man telling stories. It was as if he anticipated some doubt on the part of the readers. Yet our preliminary research has verified the names, the years, and the events he recounts with an amazing accuracy. Many of the individuals to whom he refers appear prominently in other records of the times: Capitán Mariano Tafolla; Lorenzo, Santiago's brother; Mr. R. A. Matthews and The National Hotel; Rev. Hinton; Major Van Dorn; Fred Metzger; José Policarpio Rodríguez.

We also studied the context in which he wrote in 1908, and the fact that A. J. Sowell had been documenting the lives of "Texas Indian Fighters" for several years before his book was published in Austin in 1900. And yet, among his many, many chapters on "Texas Indian Fighters" he

Santiago with sons and son-in-law, after the funeral of Anita Tafolla Mejía. (Left to right) Pete Tafolla, Santiago Tafolla, Sr., J. J. Mercado, Santiago Tafolla, Jr., Ernesto Tafolla, Lolo Mejía, Mariano F. Tafolla, Lott Tafolla, and Gabriel Tafolla.

clearly excludes all *mexicotejanos* from his chapter titles. Sowell titles more than thirty chapters after Texas Indian fighters, focusing on Dr. J. C. Nowlin, Capt. Fred Metzger, Mr. James Hudspeth, and dozens of others, and does give an introductory nod to a few of the "early Spanish settlers." But except for a few minute and casual mentions, among them the Presidentially recognized J. P. Rodríguez and Santiago himself under other names, as mere props in the chapters about other "heroes," the involvement of *mexicotejano* heroes is clearly ignored. The word "Mexican" appears only indistinctly and namelessly, as backdrops often do. ("Between Elm Creek and Medina they [the Indians] killed a Mexican boy" 118.) Is it possible that given the publication of Policarpio's book in 1898, and the local recognition and deference that all of those who had been around for the actual fights on the frontier appropriately afforded "Polly", that Sowell was forced to record the tales that Nowlin and Metzger told of "Polly Rodrigues as guide and trailer" and also of "a Mexican named Jim Tafoyia" (697) and "Jim Lafoyo, bugler"? (778). All of these occurrences, along with the frequent prejudices of the time period, could have urged Santiago to provide as many verifiable details as possible. Perhaps Sowell's effective exclusion of their stories in his preliminary research for his book was responsible for Policarpio's memoir as well.

It demands a moment of pause, even more than thirty years later, as he tries to describe the changes which have brought him to this point in his life. Interrupting his writing, and perhaps immersed in thought, he leaves the tablet, and continues with the many tasks required of his life as father, husband, grandfather, and circuit-riding preacher.

Then, in 1911, his daughter Anita Tafolla Mejía, dies one month after giving birth to her first child. Later that year, his son, Rev. Ernesto Tafolla, dies suddenly. Two weeks to the day after Ernesto's death, while Santiago is preaching to members of the Conferencia Trimestral, at the same pulpit in Seguin at which he had been baptized 35 years before, he drops dead of an apparent heart attack, at age 74. According to the ministerial report, there was no illness other than sadness because of his son's recent death.

On Oct 30, 1911, a funeral was held at the Templo Metodista Mexicana in San Antonio, at the corner of Pecos & San Fernando. Pall Bearers included Rev. Policarpio Rodríguez, Rev. Santos Romo, Rev. Pedro Grado, Rev. C. A. Rodríguez, Rev. A. R. Cárdenas, Rev. A. C. Riojas, and six more honorary bearers, including Dr. J. E. E. Harrison and Dr. V.

A. Godby. Rev. David W. Carter and the pastor, Rev. Bacilio Soto, officiated. Numerous family members stated that he is buried in the San Fernando City Cemetery of San Antonio.

His widow, Anastacia, survived him by 24 years, living at the home of her son Rev. Mariano F. Tafolla (grandfather of the editors, Carmen and Laura, and father of Fidel, the translator of this document) and telling many of the stories of the family's history to her grandchildren. On November 5, 1935, Anastacia died peacefully at the age of 93 in San Antonio, Texas. She is buried in the Herrera Cemetery, a Texas Historical site, south of San Antonio, next to the same Herrera home to which she and Santiago had been taken when ill from malaria shortly after their wedding.

The year following Santiago's death, 1912, his native New Mexico achieved U. S. statehood. Having originally joined the U. S. Army in order to return to New Mexico, Santiago died without ever having set foot in Santa Fe or seen any of his original family again. He left as his legacy, though, 14 children, more than 50 grandchildren, many devoted friends, and a thirty-five-year career as a preacher in his adopted homeland of Texas. Three of his children became ministers, one became a judge, one a *vaquero*/professional cowboy/policeman, and many were teachers. His grandchildren included judges, lawyers, teachers, one school principal, military veterans, business-owners, and ranchers, and among his great-grandchildren were a former national President of LULAC, a well-respected sheriff of modern-day San Antonio, two Ph.D.'s in Education, and at least one Civil War Camel Crew Re-enactment Specialist. One of the first Methodist preachers of Hispanic descent in the United States, his impact on the lives of many, including his own cousin-by-marriage, Policarpio, is impressive.

In the 1960s, Fidel L. Tafolla, the oldest living grandson of Santiago, who had known him and heard his stories as a child, began to explore the possibilities of publishing Santiago's autobiography, long preserved by Santiago, Jr. in a steel vault. In 1968, having recently retired from a fifty-year-career in education, he dedicated himself wholeheartedly to the document's translation to English, and had both English and Spanish versions transcribed. He contacted numerous historians and family members, researched the information at the Santa Fe Archives, and explored many possible vehicles for publication, including the Readers' Digest. He tried several times to establish a line of communication with Santa Fe archivist and historian Fray Angélico Chávez (who, as a devout

Catholic, seemed quite unenthusiastic about facilitating research on a Protestant minister's New Mexican roots. Great-granddaughter Carmen Tafolla, researching the same information in the same archives in the mid-1980s, found a much more helpful situation.)

Seeing that segments of Policarpio's autobiography had been published in Old West magazine, Fidel tried to pitch Santiago's tale as a western and submit it to publishers of novels about the Old West. He titled it "Nearing the End of the Trail" and inserted a brief introduction stating:

> The true life story of a self-made man, this account depicts the hardships of his early life and gives a faithful description of his military career touching incidentally on life in the United States before the Civil War and the life of Texas pioneers. Taken from a manuscript written in Spanish in 1908 and translated into the English language by one of his grandsons, the story has a great deal of pathos and local color and has special appeal for any student of Texas pioneer life.
>
> —Fidel L. Tafolla, 1970

But the presses of the early 1970s did not seem interested in or accepting of the possibility of a Mexican-American cowboy, especially a Protestant one. Presumably, it simply did not fit their images or their stereotypes. And indeed, Santiago's life was NOT standard, nor did he fit easily into the categories assumed at that time to be "typical" of Latinos in the United States. Fidel died of a heart attack on January 10, 1971, before he could see this book or its translation published.

In the mid-1970s, one of Santiago's grandsons, Oscar, gave his cousin's child, Carmen Tafolla, some original documents related to Santiago's life. These included, among many items, U. S. Army discharge papers from 1860, a marriage certificate from 1869, and letters of correspondence with the War Department and the Bureau of Pensions. Carmen made sure that copies of these documents, along with a copy of Fidel's English translation and other early family records and photographs, were placed with her own Tafolla Archives at the University of Texas Nettie Lee Benson Library. In the mid-1980s she traveled to Santa Fe, to verify and corroborate Santiago's story, and to begin the process of preparing the manuscript for publication, in communication with Fidel's family. Most recently, Fidel's daughter Laura found the original pencil-on-tablet manuscript handwritten by Santiago more than 100

BURIAL OF REV. J. TAFOLLA

Venerable Minister Who Fell Dead in Pulpit Will Be Laid to Rest Today.

REV. JAMES TAFOLLA.

The funeral of Rev. James Tafolla, the aged Mexican Methodist Mission preacher who dropped dead while delivering a sermon at Seguin Sunday, will be held at 9:30 o'clock this morning from the family residence, 1012 West Houston Street. Burial will be made in City Cemetery No. 5, there first being services at the Mexican Methodist Episcopal church South, San Fernando and Pecos Streets.

The active pallbearers are all men who have taken part in the Mexican Methodist mission work that has been done in South Texas and Northern Mexico, and who have known Dr. Tafolla for years. They are: Rev. J. P. Rodrigues of Floresville, Rev. C. A. Rodriguez of San Marcos, Rev. Santos Rorae of Pearsall, Rev. P. Grado of Laredo, Rev. A. R. Cardenas of C. P. Diaz, Mexico, and Rev. A. C. Riojos of Del Rio.

The honorary pallbearers are all Methodist ministers of this city. They are: Rev. S. H. C. Burgin, Rev. V. A. Godbey, Rev. J. E. Harrison, Rev. J. B. Young, Rev. C. B. Cross and Rev. Z. V. Liles.

Rev. James Tafolla was one of the pioneers and Indian fighters of this State. He enlisted in the Second Texas Cavalry, United States Army at Baltimore, Md., in 1865, and came to Texas the same year as the regiment. In 1860 he settled in Bandera County, where he resided until 1876, in the latter part of which he came to San Antonio and engaged in missionary work.

In March of 1877 he went by land in a two-horse wagon, with his family, to Laredo; prior to railroad connection south. The Southern Pacific had just entered San Antonio from the east. This was his first official appointment by the West Texas Conference, Methodist Episcopal Church South, of which he was a member until the organization of the Mexican border mission conference in this city in 1885.

An incident worthy of especial notice in his career is that, not having the necessary means to supply himself with provisions and the necessary equipment for his trip to Laredo, he mortgaged his ranch in Bandera County to secure the funds, as no appropriation had yet been made by the conference for that purpose. For several years in succession he filled some of the most important pastoral charges in the conference, having served also in the capacity of presiding elder in various districts. On two occasions he was elected by his conference clerical delegate to the general conference.

The veteran minister was always found faithfully at his post and in spite of the most inclement weather he always filled his appointments. The recent sad death of one of his sons, Rev. Ernest A. Tafolla, was a source of deep affliction to the aged father, but notwithstanding this, immediately after this sad occurrence he left for his work and was delivering the Word of God to his congregation when his death came.

Rev. Dr. Tafolla is survived by his wife, one daughter, Mrs. M. T. Mendoz; six sons, John, Petery, Gartial, James, Marion and Lott, and forty-eight grandchildren and great grandchildren, all residing in this city with one exception, John Tafolla, who is a professor in the high school at Houston.

DROPS DEAD WHILE PREACHING IN PULPIT

Seguin Minister Succumbs to Attack of Heart Failure in Presence of His Congregation.

Rev. James Tafolla, aged 74 years, a pioneer in Mexican Methodist church mission work in Texas, dropped dead in his pulpit while delivering a sermon to his congregation in his church at Seguin.

The body was brought to the city yesterday, and interment will be made in one of the city cemeteries. The funeral arrangements will be announced later.

While advanced in years, Dr. Tafolla was apparently in the best of health and spirits. His death came as a shock to his many friends in San Antonio and Seguin. Although no reason is advanced for the cause of his death, it is thought that the death of his son, which occurred a few weeks ago, was so great a shock his heart was affected.

Dr. Tafolla came to San Antonio thirty-eight years ago, and until the time of his death was known as an active and vigorous worker in his chosen profession.

For the purpose of completing arrangements for taking his wife and daughter to Seguin to make their home, Dr. Tafolla was expected to arrive in the city today.

Obituaries from San Antonio newspapers October 31, 1911, about Santiago's death.

years ago among her father's papers. Carmen and Laura set about to verify Fidel's transcription of the Spanish-language memoir against the handwritten original, in order to preserve the language and the flavor of Santiago's original Spanish-language account.

It is most appropriate that his memoir, as we know it, end with a question mark. This memoir, its epilogue, and the historical footnotes accompanying the English translation do not by any means tell the whole story, but they are a start, an invitation, a ripe field of historical material, where the spirited horses of courage, integrity, faith, and a hunger for learning run, leaving indelible hoof prints that force us to seek those unseen connections which bind us all in this borderless human experience.

—Carmen Tafolla
San Antonio, Texas

NOTES

[1] Galisteo is approximately twenty miles south of Santa Fe, New Mexico.

[2] Fidel in his notes, states that the Mexican army used bows and arrows at that time. Family stories told by Santiago to his oldest son, Pete, who repeated them to Santiago's grandchildren, among them Mariano Tafolla, (father of Carmen) and Sarah Tafolla, state that some villagers saw the soldiers spending gold coins after the incident. Tafolla, Sarah. Personal interview conducted by Carmen Tafolla. July 18 1964. Tafolla, Sarah. Personal interview conducted by Carmen Tafolla. August 2 1967. Tafolla, Sarah. Personal interview conducted by Carmen Tafolla. June 21 1972. Tafolla, Mariano Jr. Personal interview conducted by Carmen Tafolla. October 20 1972. Tafolla, Mariano Jr. Personal interview conducted by Carmen Tafolla. December 22 1972. Tafolla, Mariano Jr. Personal interview conducted by Carmen Tafolla. December 27 1972.

[3] The Mexican Archives of New Mexico state that on November 3, 1837, José Tafolla (having enlisted as José Mariano Tafolla) died "en manos de los enemigos" and they deliberated on the payment of 260 reales a month (65 pesos) to the family. This would have made Santiago three months old at the time of his father's death, but to an eleven-year-old concerned with daily mistreatment in the present, three months old could have been easily remembered as three years old. "Presidio Military Records Dec. 27, 1837." State Records Center. Santa Fe, New Mexico, accessed 1970.

[4] Rafael was also a soldier, and indeed many of the males in the community seemed to be soldier-farmers, serving at the Presidio when there was need. Military records in the Santa Fe Spanish Archives show that Mariano, Santiago's father, was a first corporal and that Lorenzo was a rifleman. Spanish Archives of New Mexico, Presidio Rolls, 1816-1824, passim.

[5] Although it is not known where El Realito, the Tafollas' ranch, was located, according to New Mexico State Senior Archivist, Samuel Sisneros, it is quite

possible that what was then referred to as El Realito, after the Gold Rush, el Real de Oro is now known as Golden, New Mexico. San Francisco Catholic Church, located there, is said to have been built around 1830, seven years before Santiago was born.

[6] The State Archives of New Mexico lists a baptismal certificate for the daughter of Santiago's brother Lorenzo Tafoya and Ana María Bael. Born October 10, 1837, and named María, her uncle Rafael was her godfather. This could have been María Francisca, and could have been the same child to whom Santiago refers, which would have made her two and a half months younger than Santiago.

The alternate spellings of Tafoya and Tafolla are not significant in a time period when it was the pronunciation, not the spelling, of a name which held meaning, and one frequently finds the "y" and "ll" spellings changed at the Presidio's roll call each time a different officer is calling military roll, so that Analla, Botello, Tafolla and Montolla might be the chosen spellings for soldiers on one day, and a different officer would record the same individuals as Anaya, Boteyo, Tafoya and Montoya on a different day. Lorenzo and most of the Tafoyas, Montoyas and Anayas currently in New Mexico used the "y" spelling, while most of Santiago's descendants (and numerous other Tafollas in the Spanish-speaking world) use the "ll" spelling. Spanish Archives of New Mexico, Reel #16. Military Presidio roll calls of Santa Fe, Spanish Archives of New Mexico, 1800-1848, passim.

[7] The original states "imperial." See notes on Santiago's later usage of this term for fabrics in Footnote 94.

[8] Las Vegas, New Mexico is approximately thirty miles due east of Santa Fe, NM.

[9] The Santa Fe Trail covers some 900 miles and traverses New Mexico, Colorado, Kansas, Missouri, Oklahoma and Texas. The Santa Fe Trail was a commercial highway connecting Missouri and Santa Fe, NM. Used by Mexican and American traders from 1821 until 1846, the Army of the West followed the Santa Fe Trail in 1846 to invade New Mexico during the Mexican-American War. When the Treaty of Guadalupe Hidalgo ended the war in 1848, the Santa Fe Trail became a national road connecting the United States to the new southwest territories. It was used by commercial freighters, military freighters supplying the southwestern forts, stagecoach lines, thousands of gold seekers heading to the California and Colorado gold fields, adventurers, fur trappers and immigrants. In 1880, when the railroad reached Santa Fe, the trail faded from importance. "Santa Fe National Historic Trail" 20 Jun. 2007. *U.S. National Park Service. 11 Aug. 2007* <www.nps.gov/safe/historyculture/index.htm>.

[10] The "Staked Plains" or "Llano estacado" is an area which covers parts of the western panhandles of Texas and Oklahoma as well as the eastern border of New Mexico. It is a high and dry area covered by short and midlevel prairie grasses and low shrubs. "Pecos and Staked Plains, Physiographic Area 55." *Bureau of Land Management*. 15 Jul. 2008 <www.blm.gov=wildlife=pl_55 sum.htm>.

[11]Original manuscript says "varas". A vara was a "rod," a standard Spanish colonial measurement, roughly equivalent to a yard.

[12]It is most certain that Plácido had heard many tales of the *americanos* against whom Mexican-American soldiers had been at war until February 1848. As a presidio town, in which a large percentage of the adult males were employed as soldiers, Santa Fe was surely filled with many reports about the war and the *americano* enemies. The Mexican-American War had ended a mere five months before this encounter with the wagon train of Americans.

[13]Original says "espías" (spies) but the more common English reference to them would be "scouts" or "guides."

[14]Council Grove, Kansas

[15]"No, sir, nothing has happened to the young lady."

[16]The Mississippi River runs through the middle of St. Louis, Missouri.

[17]One-fourth black or "quadroon" in the old Spanish caste system.

[18]The *Biographical Directory of the US Congress, 1771-present* lists a Thomas Cutting Love (1789-1853) as having been elected to the 24th Congress in New York from the 32nd District for the years 1835-1837. He died in Buffalo, New York, in 1853. It is possible that this was the brother of Mr. Love whom Santiago mentions. There is also a Rep. Peter Early Love, native of Dublin, Georgia, who was a U.S. Representative from Georgia (March 1859 to January 1861) but it is unlikely that he was the brother mentioned because he would not havc yct bccn clcctcd at thc timc of Santiago's visit.

[19]Franklin Pierce was President of the United States (1853-1857).

[20]General Winfield Scott led several brilliant attacks in the Mexican-American War. It is said (by American sources) that General Scott so impressed some of the Mexican people that a delegation later visited his camp to inquire as to the possibility of him leading their new government. Intent on becoming president of the United States, he declined. On February 2, 1848, the signing of the Treaty of Guadalupe Hidalgo ended the two year Mexican-American War. The government of Mexico ceded 525,000 square miles (55% of its holdings, not including Texas) which included parts of modern day Colorado, Arizona and Wyoming, as well as all of California, New Mexico, Utah and Nevada in exchange for 15 million dollars (about $313 million circa 2006) as compensation for war-related damages to Mexican land. In addition Mexico agreed to recognize the Rio Grande River as the border between Texas and Mexico, and the U.S. promised to protect the property and civil rights for former Mexican citizens now living on U.S. soil. "General Winfield Scott and the Mexican-American War 1846-1848." *About North Georgia.* 11 Nov 2008 <http://ngeorgia.com/other/scottinmexico.html>. "Treaty of Guadalupe Hidalgo." 23 Jun. 2009. Wikipedia. 23 Jun. 2009 <http://en.wikipedia.org/wiki/Treaty_of_Guadalupe_Hidalgo>.

[21]John Russell Bartlett, an 1800s student of American history and linguistics, who is best known for having published an 1848 *Dictionary of Americanisms*, mentions R. A. Matthews in his *A Personal Narrative of Explorations and Incidents in Texas, New Mexico, California, Sonora and Chihuahua*, 1854, (Vol. II, Appendix C.) as an assistant in the surveying, astronomical and topographical division of the field party organized in Washington August 1850 for field duty to survey the new United States and Mexican boundary. Charles Radzinski was listed as a Principal Assistant Surveyor on p. 400 of the same volume. (It is most likely that this was the Charles Radziminski, popular 1st Lt. of Company K, Second Cavalry, who died of tuberculosis c.1858 and for whom was named the new post, Camp Radziminski, of the Wichita Expedition in which Santiago served.) Matthews' work with Congress in Washington, D.C. also suggests that he may have served as an information specialist and advisor to congressmen dealing with the new border issues. Of Matthews' two companions mentioned on the trip to Washington, Love and McGoffin, we find in Bartlett's report a James W. Magoffin, Esq., a long-time resident of Mexico who by October of 1852 had settled into a rancho on the site of present-day El Paso, and who enthusiastically invested in the border area. He graciously hosted the Bartlett mission. In addition, Kendall's *Narrative of the Texan Santa Fe Expedition*, describing a trek from Texas to New Mexico a decade earlier, mentions a Mr. Samuel Magoffin, "a native of the United States, but at this time a merchant of Chihuahua who was now on his way to that city with more than 40 wagons heavily laden with goods." This Magoffin visited Kendall in Albuquerque when he and other Americans were being held captive there by the Mexican officials, and brought them "a generous supply of coffee and tobacco." It is likely that the two were related, or the same person. John Russell Bartlett. *A Personal Narrative of Explorations and Incidents in Texas, New Mexico, California, Sonora and Chihuahua 1850-1853*. Vol. II, Appendix C. 383-384, 400-402. Print. George Wilkins Kendall. *Narrative of the Texan Santa Fe Expedition*. 444. Print

[22]The National Hotel was located at Pennsylvania Avenue and 6th Street, several blocks from the White House. President Lincoln would later give a speech from its balcony. It was also the hotel preferred by actor John Wilkes Booth on his trips to Washington, and he would be a guest there in the days before he assassinated President Lincoln in 1865. "The National Hotel." *Mr. Lincoln's White House*. 18 Jan. 2009 <www.mrlincolnswhitehouse.org/inside.asp?ID=185&subjectID=4>.

[23]It is unknown what the term "majoros" meant to Santiago, although it is possible that he was employing a colloquial or phonetically altered usage of "masoros," meaning those who study the Masorah, a reference to the Hebrew work on the Bible used to preserve the text from corruption. The term could have possibly, colloquially, come to mean those who study religious texts. One other possibility is that he was referring to "masones" (Masons or Freemasons) although it is unknown whether they existed publicly in New Mexico at

that time, as the Catholic Church was very anti-Masonic, with Papal declarations against them from 1738 to 1917, and more currently, open Anti-Masonic statements by then-Cardinal Ratzinger, now Pope Benedict.

[24]"Layman" here probably refers to a non-clergy teacher. Often religious brothers ran Catholic schools, and would have been distinguished from the priests as "laymen."

[25]A *colegio*, often translated college, is a private school, either K-12 or university level.

[26]In the original manuscript, Santiago spells it "Talbaton." Talbotton is some 35 miles from Columbus, GA, and the site of the Methodist Institute to which he refers as the Talbaton Institute.

[27]Muscogee County is located in Georgia.

[28]The Muscogee Railroad was a 50-mile railroad running east from Columbus to Butler, GA. Storey, Steve. "Muscogee Railroad." *Georgia's Railroad History & Heritage*. 19 Dec. 2008 <http://railga.com/musco.html>.

[29]Neil Foley in his brilliant study of perceptions of Mexicans and Blacks in the nineteenth-century South notes the ambivalence of White attitudes toward Mexicans, depending on the situation's potential for maintaining their status over Blacks. He quotes Frederick Law Olmsted's criticism of the Antebellum "slavocracy" where to "the vulgar mind of the South" Mexicans are "considered to be heathen; not acknowledged as 'white folks'" and the White Texan dislike for Mexicans' openness to intermarriage with Blacks and their tendency to "associate and sympathize with the Negroes." Foley states that Mexicans became a threat for "crossing the color line as well as the border in aiding slaves to freedom in Mexico" and contrasts the treatment Santiago received in Georgia with treatment he would later receive in Texas. Again, he quotes Olmsted in 1857, "wherever Mexicans came into contact with slaves in antebellum Texas, 'it has been found necessary to treat them as outlaws.'" Commenting on Santiago's interesting intercultural status, he calls the young boy (who was between 14 and 16 years of age at the time) "perhaps history's only Mexican overseer of a slave plantation in the antebellum South" (23). Foley, Neil. *White Scourge: Mexicans, Blacks and Poor Whites in Texas Cotton Culture.* Berkeley: U of California P, 1998. Olmsted, Frederick L. *A Journey Through Texas*. Austin: U of Texas P, 1978.

[30]A school named the Collinsworth Institute was located in Talbotton, Georgia. (An early Georgia newspaper, *The Columbus Enquirer*, notes the death of "Rev. John Collinsworth, 51, an itinerant minister of the Methodist-Episcopal Church" in their Sept. 13, 1834 issue, and it is possible that this was the individual for whom the institute was named.) Established in 1837 as a manual labor school for boys, the early concept was not very successful, and it later became a college preparatory school for Methodist boys. H. H. McQueen was an instructor from 1841-1856. During that time period, he bought the school from stockholders and consequently sold it in 1856. After the war, the school was reinstated. One of the

board members was Judge J. M. Matthews. It is unknown whether Judge Matthews was the son of F. T. Matthews and brother of R. A. Matthews. There is also a Rev. W. D. Matthews mentioned in a May 3, 1837 issue of the same paper who performed a marriage ceremony, and it is very possible that this individual was related to the Matthews family Santiago knew. But it appears likely that this was the same "Talbaton Institute" (his spelling) which Santiago attended circa 1851. The New Georgia Encyclopedia—Cities and Counties, 19 Feb. 2009 <http://www.georgiaencyclopedia.org/nge/Article.jsp?id=h-2393>. <http://files.usgwarchives.net/ga/talbot/history/collings.txt>. The Columbus Enquirer, Sept 13, 1834; Columbus Enquirer, May 3, 1837.

[31]In 1833 the Georgia Methodist Conference envisioned the creation of a church-sponsored school which would allow students to do farm work and also prepare for college. In 1834 The Georgia Methodist Conference Manual Labor School was chartered. This school would later be identified as Emory College (1840) and subsequently Emory University. In 1848, Joseph S. Key graduated from Emory University. Thirty-eight years later he became a bishop in the Methodist Church. When Santiago refers to Bishop Key attending the same school as he, it is not clear whether he was referring to this school or to the Collinsworth Institute, or whether they were interconnected as part of the Methodist educational system. "A Sesquicentennial Timeline: 1833-1987." 2008. *Emory University.* 13 Aug. 2008 <http://emoryhistory.emory.edu/timeline/index.html>.

[32]One preacher mentioned in several Methodist church services in Georgia in the latter half of the nineteenth century was The Rev. J. W. Hinton. *The Talbotton New Era* newspaper on Tuesday, January 5, 1897 reports, "Dr. J. W. Hinton, the new presiding elder of the Columbus district, occupied the pulpit at the Methodist church Sunday morning and evening. He preached excellent sermons on both occasions. Thirty-nine years ago he came on this district as presiding elder. . . . Sunday was the anniversary of his birthday, which anniversary he did not reveal to his audience but he told them that he began to preach at the age of 20 and has been preaching 52 years." This would have given him a birthdate of January 3, 1825, and would have made him about 25 years old at the time Santiago met him, if this is indeed the same Rev. Hinton.

[33]Don José Manuel Gallegos was educated by Franciscan missionaries in Taos, New Mexico, and became a Catholic priest at the age of twenty-five. As pastor of San Felipe de Neri Church in La Villa de Albuquerque, he became politically involved, becoming an elector of representatives to the Mexican Congress. After the Treaty of Guadalupe Hidalgo made New Mexico a U. S. territory, Don José Manuel Gallegos was elected to its first Territorial Council. In 1853, he became the first Democrat elected as a representative to Congress from the Territory of New Mexico. As such, Gallegos became only the second Hispanic in U. S. history to have been elected as a Congressional representative, the first being Joseph Marion Hernández from the Florida Whig Territory in 1822. "Don

José Manuel Gallegos." 29 Oct. 2007. *The Library of Congress*. 11 Nov. 2008 <http://lcweb2.loc.gov/ammem/today/oct30.html>.

[34]Major José D. Sena was born in Santa Fe in 1837. He was engaged in service during the Civil War and, at the end of it, was in charge of the rebuilding of Ft. Marcy in Santa Fe. In addition to Sheriff of Santa Fe, he held many offices of trust including interpreter in the courts. He was admitted to the bar in Virginia and held the office of register of the U. S. land office. Twitchell, Ralph Emerson. *The Leading Facts of New Mexican History Vol. 2*. Cedar Rapids, Iowa: The Torch Press, 1912. Print.

[35]Jefferson Davis, (1808-1889) was a Mississippi congressman in 1845, a U. S. senator 1847-51 and 1857-61, and President Pierce's Secretary of War from 1853 to 1857. The newspaper Santiago saw, which told of recruits being sought, must have been one of its earliest announcements, for it was not until March 3, 1855 that Congress passed a law authorizing two mounted regiments that would be known as cavalry. They would be the first in American military history to bear that name. By March 26 Davis established the regiments and had appointed the officers. It must have been about March 30 when Santiago made his visit to Secretary of War Jefferson Davis' office, and April 1st when he enlisted. This timeline reveals the eagerness of both Jefferson Davis and Santiago Tafolla, for different reasons. The intersections of Davis' life with Santiago's appear twice more after this. It was Jefferson Davis who espoused the project to bring camels to the Southwest, and Santiago was employed twice with the camel corps. Then, in 1861, Davis was chosen by a provisional congress to serve as President of the Confederate States of America, and remained in that role until the end of the Civil War. Santiago was to serve in the Confederate Army from 1862 to 1864. In 1865 when the war ended, Davis was imprisoned by the U. S. government for treason (based on his Confederate service), but after two years in prison was released on bail, after which, charges were dropped. "Jefferson Davis 1808-1889 President of the Confederate States of America." 9 Jul. 2009 <http://americancivilwar.com/south/jeff davi.html>.

[36]Fort McHenry is a star-shaped fort known for its role in the War of 1812. When American soldiers successfully defended Baltimore Harbor from an attack by the British navy in the Chesapeake Bay, Francis Scott Key witnessed the survival of the U. S. flag during bombardment and was inspired to write *The Star-Spangled Banner*, the poem that would eventually be turned into the national anthem of the United States. <http://www.nps.gov/fomc/historyculture/index.htm>.

[37]Jefferson Barracks is in St. Louis, Missouri, and was built near the confluence of the Mississippi and Missouri Rivers. It was the largest military reservation of its day, encompassing over 2000 acres and stretching about two miles along the Mississippi River. We know that the recruits for the Second Cavalry came from Alabama, Indiana, Kentucky, Ohio, Maryland, Missouri, and Virginia, and that the horses were assigned to companies by color: grays for Company A, sorrels

for B and E, bays for C, D, F, and I, browns for G and H, and roans for Company K, so we know that Santiago would have been riding a bay. Drummond, Loren. "St. Louis County Parks & Recreation: Jefferson Barracks Historic Park." 23 Jun. 2002 <http://www.civilwaralbum.com/misc/jeffbarracks1.htm>.

[38]This is the same Robert E. Lee who later served as Commander in Chief of the Confederate forces in the Civil War.

[39]The original manuscript refers to Van Dorn consistently as Vandon, but given the Southern accents they would be using, this would probably have been pronounced Van Dawn, and Santiago appears to have been writing it down phonetically. Most of the names he records here are fairly accurate, with some spelling alterations, such as Oats for Oakes and Stone for Stoneman, and a few company commanders misplaced. In response to Indian attacks, the Second U. S. Cavalry was organized specifically to protect the Texas frontier. In late 1855, Col. Albert S. Johnston, Lt. Col. Robert E. Lee, and Major John Bell Hood brought some 750 cavalrymen to headquarters at Ft. Mason, TX, and the nearby outpost of Camp Verde, TX. The Second Cavalry included sixteen future Civil War generals including: Robert E. Lee, Albert Sidney Johnston, William J. Hardee, Edmund Kirby Smith, John Bell Hood, and George Thomas. In addition, Captain George Stoneman, whom he refers to here, became the highest ranking Union officer captured during the Civil War. "Historic Camp Verde." 23 Jun. 2007. *Kerr County Historical Commission.* 23 Jun. 2007 <http://www.co.kerr.tx.us/historical/camp_verde.htm>. "Timeline of Texas and the Western Frontier 1856-1865." 30 Jun. 2003. *Texas Beyond History. 2 June 2009* <http://www.texasbeyondhistory.net/forts/56-65.html>.

[40]Fidel's transcription adds "y Creeks." Whether this reflects some story he heard as a child or something he researched is unknown. "Creeks" is not written in Santiago's original manuscript.

[41]Santiago is amazingly accurate in his recollection of dates. Military records verify that it was, indeed, October 27, 1855 when the regiment departed Jefferson Barracks bound for Texas, that they crossed the Red River on December 14 and arrived at Fort Belknap on December 27,1855. Chalfant, William Y. *Without Quarter: The Wichita Expedition and the Fight on Crooked Creek.* London: U of Oklahoma P, 1991. Print.

[42]The "Texas Northers" are renowned for their quick drops in temperature, and Johnston's own letters emphasize this, describing a norther that lasted more than five days and featured below-zero temperatures and 60 m.p.h. winds. Johnston, William Preston. *The Life Of General Albert Sidney Johnston: Embracing His Services In The Armies Of The United States, The Republic Of Texas, And The Confederate States.* New York: Da Capo P, 1997. Print. (Johnston,188)

[43]In this first mention in the manuscript, Santiago spells the name "Prusia" here, but in later references to him, Santiago spells the name "Prusa."

[44]It is unclear whether these were wild boars or domesticated pigs from the farm.

[45]The Alamo was a Spanish mission originally named San Antonio de Valero and was one of five missions located in the San Antonio area in the early 1700s. The mission was secularized and became a cavalry unit for the Spanish military in the early 1800s. In 1805 the abandoned convent portion of the Alamo became the first military hospital in Texas. In 1836, during the Texan revolution against Mexico, a band of some one hundred eighty-seven Texans and Tejanos, under the command of William B. Travis and James Bowie, chose to keep and defend their military post in San Antonio, against the orders of General Sam Houston, who had been advised by fellow revolutionary Juan Seguín that the entire Mexican Army was marching on San Antonio. Seguín had sent his nephew, Blas Herrera, to gather information on the Mexican Army. While many San Antonio families, hearing of Santa Ana's march on San Antonio, evacuated their families from the city, Travis and his men chose the Alamo as their defense point. The thirteen-day Mexican siege of the Alamo has drawn much historical attention, and the fall of the Alamo on the morning of March 6 resulted in the battle cry "Remember the Alamo" during the subsequent battle of San Jacinto, in which Santa Ana was captured and Texas won its independence from Mexico. The Blas Herrera, emissary of the information to aid the Texan forces, was the father of Blas Herrera, Jr., who married Sabinita, Anastasia's sister, and whose residence Santiago mentions several times after his marriage to Anastasia. The two Blas Herreras (Sr. and Jr.) as well as Sabinita Salinas Herrera and Anastacia Salinas Tafolla were buried in the Herrera Cemetery (to the side of the historical Herrera residence, no longer standing), which was declared an official Texas Historical Site. Santos, Richard. *San Antonio de Bexar—1 de Enero 1836.* San Antonio, 1963. Print. Ramírez, Amalia, Family documents related to Sabinita and Anastasia, 1977.

[46]*Jacales* in San Antonio at that time were huts often made of adobe or sticks and had thatched roofs made of straw, grass or palm leaves. Olmsted's *A Journey through Texas* in the 1850s describes in the square of the Alamo, "windowless cabins of stakes, plastered with mud and roofed with river-grass, or tula . . . or low, windowless, but better thatched houses of adobe . . . " Olmsted, Frederick L. *A Journey Through Texas*. Austin & London: U of Texas P, 1978, 149)

[47]The Menger Hotel in San Antonio opened on February 1, 1859, and was host to many famous people including Teddy Roosevelt, who reportedly recruited his "Rough Riders" in the Menger Bar. The Menger is still in operation today as a luxury hotel. The 150-year-old building is said to be haunted and many guests have reported encounters with the resident ghosts. "Menger Hotel San Antonio, Texas." *National Trust for Historic Preservation.* 12 Sep. 2008 <http://www.historichotels.org/hotel/the_Menger_Hotel/>.

[48]In August of 1856, Camp Verde, on Verde Creek was established as a U. S. Cavalry post. According to A. J. Sowell, it was established by an act of Congress in 1855 "for frontier protection, and as a camel post." Captain I.N.

Palmer was the first commander, followed by Major Bowman, Lt. Wheaton, Gen. Robert E. Lee, and Gen. Albert Sidney Johnston. Emmett, Chris. *Texas Camel Tales.* San Antonio: Naylor, 1932. Print. Hunter, J. Marvin. *Old Camp Verde: The home of the camels: a romantic story of Jefferson Davis' plan to use camels on the Texas frontier*. Bandera, Texas: Hunter, 1948. Print. Sowell, A. J. *Early Settlers and Indian Fighters of Texas*. Austin & London: U of Texas P, 1900. Print.

[49]It is possible that that the elderly Mormon leader to whom Santiago refers is Lyman Wight. In 1845, Mormon leader Lyman Wight's unconventional views and contrary nature had caused him to fall from his leadership position in the established Mormon Church in Utah. Wight and a group of some 200 Mormon followers came to Texas, eventually settling in Gillespie County where they established the community of Zodiac. The Mormons built a sawmill, a grist-mill, a school, a temple, and a store, and became valuable members of the community of Fredericksburg, as they supplied the German settlers there with seeds, lumber, and flour. After floods destroyed their mills, the Mormons left Gillespie County and settled briefly in Burnet County, before moving to Bandera County in March of 1854. In the fall of 1856, Wight and his followers established a settlement which came to be known as Mormon Camp in Mountain Valley on the Medina River. On September 8, 1856, Wight performed the marriage of his son Levi Lamoni to Sophia Leyland. The Mormons are also mentioned by Williams, who states that they "had established themselves on the Medina at the time that the main body of their curious co-religionists were settled in Nauvoo." Banks, Stanley C. "The Mormon Migration Into Texas." *Southwest Historical Quarterly* Oct. October 1945: 49. Wight, Levi Lamoni. *The Reminiscences and Civil War Letters of Levi Lamoni Wight.* Salt Lake City: U of Utah P, 1970. Print. Gillespie County Historical Society, *Pioneers in God's Hills* (2 vols., Austin: Von Boeckmann-Jones, 1960, 1974). Taylor, T. U. "Lyman Wight and the Mormons in Texas." *Frontier Times.* June 1941. Williams, R.H. *With the Border Ruffians: Memories of the Far West, 1852–1868.* New York: Dutton, 1907. Print.

[50]"For I am a man under authority, having soldiers under me: and I say to this *man*, Go, and he goeth; and to another, Come, and he cometh; and to my servant, Do this, and he doeth *it*." Matthew 8:9 (King James Version.)

[51]After Texas became a state in 1845, settlers coming in from other states faced many challenges including long distances, rugged dry terrain, severe weather, and attacks by Indians and outlaws. In response to these harsh and often hostile conditions on the Texas frontier, Secretary of War Jefferson Davis proposed an experiment to bring camels to the United States. It was thought that the camels would be able to withstand extreme conditions better than horses and mules. Congress agreed and passed the "Camel Appropriation Bill." In 1856, thirty-three camels arrived at the Port of Indianola Texas and were led to their permanent location at Camp Verde, Texas. By the time the Confeder-

ate Army overtook the camp in 1861, there were reportedly eighty camels stabled there. The Confederacy used them on a limited basis to transport cotton to Mexico. After the Civil War ended, the camel experiment was largely abandoned; some camels were sold at auction, others were released onto the open range to fend for themselves. Emmett, Chris. *Texas Camel Tales.* San Antonio: Naylor, 1932. Hunter, J. Marvin. *Old Camp Verde: The Home of the Camels: A Romantic Story of Jefferson Davis' Plan to Use Camels on the Texas Frontier*. Bandera, Texas: Hunter, 1948.

[52]It is possible that the captain is referring to his frequent referral of "opportunity" and responsibility to Santiago, in some recognition of his character and dependability and the fact that he could be counted on to not be a slacker. However, in terms of consideration, this is the same captain that ordered him to make useless trips, and according to Santiago's assessment, this was possibly because the captain valued his life more lightly because of his ethnicity.

[53]Santiago spells it "Nowling" in his original, but A. J. Sowell and other sources call him Dr. Nowlin. Born in Kentucky in 1817, he came to Texas in 1855 by way of Tennessee, Missouri, and Kansas. Nowlin's nephew, Henry Moore, a single man, came with his uncle to Texas, and they settled in 1856 in Kerr County near the town of Center Point. Sowell, A. J. *Early Settlers and Indian Fighters of Texas*. Austin, Texas: B. C. Jones & Co. Printers, 1900; Austin: State House Press, 1986.

[54]Nowlin is also mentioned in the *Old Camp Verde* book as living three miles north of Bandera Pass in 1856, when Camp Verde was founded with the arrival of thirty-three camels. Hunter, J. Marvin. *Old Camp Verde: The Home of the Camels: A Romantic Story of Jefferson Davis' Plan to Use Camels on the Texas Frontier*. Bandera, Texas: Hunter, 1948. Print.

[55]Known as José Policarpio Rodríguez in the first publication of his own memoir, which he dictated to Rev. D. W. Carter, who transcribed and arranged for publication by the Methodist Episcopal Church South in 1898, Policarpio became a life-long friend (and eventually pallbearer) for Santiago. Santiago calls him Policarpio throughout, and the 1850 Bexar County census lists him as "Policarpio Rodrígues" although some other sources, following the publisher's lead, say Policarpo, so we have deferred to Santiago's name for him in this book, except where Santiago himself refers to him as J. P. or his nickname among the *americanos*, "Polly." Policarpio's memoir was dictated "at odd times during the years 1892 to 1897" with some evidence of language distortion in the transcription (McCabet, Captain Palmore, etc . . .) and covers from his birth in Zaragoza, Mexico, thirty-five miles west of Eagle Pass, Texas, in 1829, up through his conversion by Santiago and to about 1880. Some of his descendents still live on his ranch on Privilege Creek, near Polly's Chapel, the rock church he built in sight of Polly's Peak, named for him. In the Presidential Unit Citation for the 2nd Cavalry, Company D, Camp Verde, presented before the 35th Congress in 1858, he is mentioned by name: "Great praise is

due to the guide, Polycarpio Rodrigues, for his untiring exertions and sagacity in the pursuit." First Sergeant McDonald of Company D is also mentioned along with "fourteen men of that company" for re-capturing horses stolen by the Indians on the San Jeronimo River. It is possible that Santiago is one of these fourteen. Rodriguez, José Policarpo. *The Old Guide: Surveyor, Scout, Hunter, Indian Fighter, Ranchman, Preacher.* Nashville & Dallas: Publishing House of the Methodist Episcopal Church South, 1898. Print. National Archives, House of Representatives, 35th Congress, 2nd Session, No.2, Presidential Unit Citation, 2nd Cavalry, Company D, Camp Verde.

[56]The original says *varas*.

[57]The original "una carabina" could refer to a carbine, a very short rifle, or to a rifle of any length.

[58]Fred Metzger of Hondo, Texas, is mentioned several times in Santiago's memoir as a good and loyal friend. Metzger is also listed in A. J. Sowell's *Early Settlers and Indian Fighters of Texas*, (Austin, Texas 1900) and his house in Hondo is listed as "the first brick home built in Hondo in 1876." Hondo Historical Houses website lists it as "More than 100 years ago, Frederick Metzger built this home and store, which became a stagecoach stop" and had "wonderful, two-foot thick walls." (Medina County History, Vol. 1, p.152, published by the Castro Colonies Area Heritage Association.)

[59]Rodríguez, José Policarpo. *The Old Guide: Surveyor, Scout, Hunter, Indian Fighter, Ranchman, Preacher*. Ibid.

[60]Dragoon revolvers were longer than normal pistols.

[61]The original says "tocaron pito" which means literally, they sounded a whistle. The story of this same fight is mentioned in two different accounts by "Indian fighters" in Sowell's book. Dr. J. C. Nowlin relates this encounter about "A Mexican named Jim Tafoyia" and also mentions Martin's wound and perforated lung, which together with a cut stomach caused him to vomit blood. And in a chapter about Fred Metzger, Sowell records an 1857 Indian fight in which "Jim Lafoyo, bugler . . . engaged an Indian at close quarters and attempted to fire at him with a pistol . . . " This chapter also states that "Lafoyo rode behind the wounded man and held him on his horse, as he was shot in the stomach and could not lie on a stretcher." However, his version of the account recounts Martin himself as having had the premonitionary dream, and Metzger who advises him to not go into battle. Apparently, Sowell did not connect the two tales as being about the same fight, and thus records them as two different incidents, and documents Santiago as two different people, with two different misspellings of his name. This same tale is also told in detail in Policarpio's book, *The Old Guide.* Sowell, A. J. *Early Settlers and Indian Fighters of Texas.* Austin, Texas: B.C. Jones & Co. Printers, 1900; Austin: State House P, 1986. Print. Rodríguez, Jose Policarpio. *The Old Guide: Surveyor, Scout, Hunter, Indian Fighter, Ranchman, Preacher.* Nashville & Dallas: Publishing House of the Methodist Episcopal Church South, 1898. Print.

[62]Some of the other descriptions of soldiers in similar battles described hides stretched between two horses, where the wounded could travel by horseback more comfortably, such as the one in which Major Earl Van Dorn travels after being wounded in late 1858, at the same battle wherein Lt. Van Camp was killed. Chalfant describes it as "a litter fastened between two horses hitched in tandem." But in Sowell's book, the reports given him by Metzger describe a different transport in which Santiago "rode behind the wounded man and held him on his horse." Chalfant, William Y. *Without Quarter: The Wichita Expedition and the Fight on Crooked Creek.* Norman & London: U of Oklahoma P, 1991. Print. Sowell, A. J. *Early Settlers and Indian Fighters of Texas.* Austin, Texas: B.C. Jones & Co. Printers, 1900; Austin: State House Press, 1986. Print.

[63]Camp Cooper was established by the Texas legislature in 1856 to protect the frontier and monitor the nearby Comanche reservation. Col. Albert Sydney Johnston was the founding chief in January 1856, and it became headquarters for four companies of the Second U. S. Cavalry under the command of Lt. Col. Robert E. Lee (his first command of a fort.) from April 9, 1856 to July 22, 1857. Captains under his command included Earl Van Dorn and Theodore O'Hara. The harsh weather and insect conditions were difficult on Lee, and he left the post in 1857 for San Antonio. In a letter to his wife in 1856, Lee, who began to keep a rattlesnake in his tent "as a pet," said the air was like "a blast from a hot furnace" and that the temperature was 112 degrees "in the Hospital Tent, the coolest place we have." (If Santiago's dates are correct, this would have been shortly before Santiago's arrival there.) Maj. George Thomas took over in 1857. Battles with the Indians included the re-capture of Cynthia Ann Parker, (mother of the famous Comanche leader Quanah Parker) who had lived and married among the Comanche, before her forced re-patriation to Anglo Texan society. Davis, Charles G. "Camp Cooper." 15 Jan. 2008. *The Handbook of Texas Online.* 12 Aug. 2008 <http://www.tshaonline.org/handbook/online/articles/CC/qbc9_print.html> Texas Almanac 2008-2009 "Camp Cooper, Fort Belknap, and the Indian reservations" 18 Jan. 2008 <http://www.fort tours.com/main.html>.

[64]The first major U.S. Army expedition against the Comanches, the Wichita Expedition was led by Captain (Brevet Major) Earl Van Dorn. Units of the Second Calvary marched north from Belknap, Texas, and established Camp Radziminski at the edge of the Wichita Mountains. Many skirmishes and two major battles ensued, the Battle at the Wichita Village and the Battle of Crooked Creek, which dealt a total blow to the inhabitants of one major Comanche village. Chalfant, William Y. *Without Quarter: The Wichita Expedition and the Fight on Crooked Creek.* Norman & London: U of Oklahoma P, 1991. Print.

[65]The editor's research for a tribal chief in this area named "Ketomsey" yielded no results. However, there is a possibility that Santiago, not speaking the language of the Indians, could have misremembered the name "Tecumseh" as

"Ketomseh" (switching the consonants), as Tecumseh was a popular name of the time period.

[66]Pohchanah-kwoheep (called Buffalo Hump by the whites) was one of the most capable war leaders of the Penateka Comanches. Maj. Earl Van Dorn, commanding U. S. troops and apparently unaware that Buffalo Hump's band had recently signed a peace treaty with the United States, attacked and killed about 80 of the Comanches at the Battle of Wichita Mountains. Buffalo Hump then led his followers to the Kiowa-Comanche reservation near Fort Cobb, Oklahoma, tried to set an example as a rancher and farmer, and died in 1870. Chalfant, William Y. *Without Quarter: The Wichita Expedition and the Fight on Crooked Creek.* Norman & London: U of Oklahoma P, 1991. Print.

[67]The great value of horses in this setting should not be underestimated. Not only were they critical to survival, but by the analysis of Robert E. Lee, Brevet Colonel, the number of horses the forces held was much lower than what they needed. "The speediest way of rendering the second cavalry more effective would be to increase the number of horses to each company, which, by . . . orders [of] . . . 1856, were reduced to sixty." *Troubles on Texas Frontier* (Thirty-Sixth Congress, First Session, Executive Document No. 81) Government Printing Office: 1860, DRT Library, Print.

[68]The reports from Camp Cooper verify that the soldiers were underfed. Chalfant documents the tendency of Van Dorn to not only permit but encourage his men to engage in hunting, sports, and any distraction to relieve the monotony of military training and the stress of the hardships of bitter cold and frequent food shortages. Chalfant, William Y. *Without Quarter: The Wichita Expedition and the Fight on Crooked Creek.* Norman & London: U of Oklahoma P, 1991. Print.

[69]Related wildcats in the area are consistent with this (*león* in the original) being a cougar, or puma.

[70]Santiago's new bride was Juanita Torres, and was probably Policarpio's cousin. Policarpio and Santiago subsequently called each other "Primo" (Cousin) throughout their lives.

[71]The descendents of Policarpio still live on the family's original holdings in that area, and hold yearly family reunions at "Polly's Chapel."

[72]In the handwritten original, the number 2 seems to have been written over the first 0, indicating 320 acres.

[73]Col. James Duff, born in Scotland in 1828, was a San Antonio merchant by 1856. Texas seceded from the Union on February 1, 1861, and Duff became commander of a Confederate military unit known as "Duff's Partisan Rangers." Later his command expanded into the 33rd Texas Cavalry. Williams, R. H. *With the Border Ruffians: Memories of the Far West, 1852–1868.* New York: Dutton, 1907; 2d ed., London: Murray, 1908. Print.

[74]It is possible that this Manuel Menchaca was a son of the Antonio Menchaca who served with the Texas Rangers, who were more mixed with Hispanics and Indians in their early days than after the 1850s. (Policarpio also mentions

Antonio Menchaca.) Placido, a Tonkawa chief, often served with the Rangers as a scout from the days of the Republic through the 1850s. However, as the tribes relocated or were removed from Texas, Indian involvement in the Rangers decreased.

During the early years of the Texan republic, Hispanics served alongside Anglo and Indian Rangers. A number of men such as Juan Seguin and Antonio Menchaca gained reputations as skilled leaders during the Texas Revolution and later used those skills in the Rangers. Some companies were composed entirely of Hispanics, but others, like those led by Lewis Sanches and José María Gonzales, were a mix of Hispanics and Anglos. Policarpio Rodríguez was also a Texas Ranger. After the war with Mexico, "strained relations between Anglos and Hispanics led to a decline in Hispanic membership in the Rangers." (From the official Texas Ranger History site of the Texas Ranger Museum.)

[75]The "Mexican Chief of Staff" Cortinas whom Santiago mentions here appears to be the same Mexican War veteran Juan N. Cortina famous for his post war defense of *mexico-tejanos* against the abuses of the new U.S. troops, and the same individual responsible for the two so-called "Cortina Wars." Involved in the Mexican struggle for independence from France and the execution of Maximilian, Cortina proclaimed himself governor of Tamaulipas in 1863 and was promoted to general of the Mexican Army of the North by President Benito Juárez. Thompson, Jerry. "Juan Nepomeceno Cortina." 17 Jan. 2008. *The Handbook of Texas Online*. 16 May 2009 <http://www.tshaonline.org/handbook/online/articles/CC/fco73.html>.

[76]Col. Santos Benavides, the highest ranking Mexican-American in the Confederacy was born in Laredo in 1823. He won distinction as a public servant, Mayor of Laredo, Chief Justice of Webb County, and leader of several campaigns against the Lipan Apaches. Much admired as a brilliant fighter, he had been offered a Union Generalship by Col. Edmund J. Davis, but accepted instead a Confederate position as Captain in the 33rd Texas Cavalry. He led many *mexicano* Confederates in successful battles against Federal forces, Mexican forces, and hostile Native American forces. Santiago spells it Benavides in his manuscript. Riley, John Denny. "Santos Benavides: His Influence on the Lower Rio Grande, 1823-1891." Diss. Texas Christian University, 1976. Print. Thompson, Jerry Don. "A Stand along the Border: Santos Benavides and the Battle for Laredo." *Civil War Times Illustrated* Aug. 1980. Print.

[77]The siege of Vicksburg began May 19, 1863, and the surrender of Vicksburg was July 4, 1863, in effect leaving the Union in control of the Mississippi River. In addition, the Confederate loss of Gettysburg was also on July 4, 1863. The fall of Atlanta happened on Sept. 2, 1864. Lee's army was plagued with food shortages and massive desertions, and the Union Army had a decisive victory on April 1, 1865. The Confederate Government officially dissolved on May 5, 1865. While news of battles and of consequences traveled slowly enough at that

time period, perhaps delaying knowledge of some of these losses for three to four months, it is not true, as has often been claimed, that the last battle in the Civil War, fought at Palmetto Ranch, Texas, on May 12-13 of 1865 (after the Confederate Surrender at Appomatox) was fought because the Confederate troops stationed there in Texas did not know the Civil War had ended. (A thesis by Jeff Hunt relates that a letter sent home from one of the soldiers at Palmetto Ranch states they had received a copy of an April 30th issue of the *New Orleans Picayune* and fought the battle because they were frightened of the bad treatment they had heard Union soldiers were giving Confederate prisoners, and the accompanying devastation of the cities.) Thus, it is likely that Santiago and the other soldiers had been receiving recent bad news from the battlefronts and had perhaps known for two months or more about Vicksburg and Gettysburg. It is the editor's belief that the instigating factor here in the desertions appears to have been not the loss of hope for the Confederacy although that was clearly perceived and surely a cause for the mass desertions by soldiers, but a fear for their lives at the direct and immediate threat of hanging by their anti-mexicano fellow Confederates.

[78]Obviously, he is referring here to that part of Mexico to which they are headed. His early years of life in Santa Fe, up through age eleven, were under the Mexican flag, and his fellow soldiers had also made incursions into Mexico during his military assignment on the U. S.-Mexican border area in Brownsville, but that area was significantly different, covering mainly Gulf and beach areas.

[79]Santiago says *frontera*, which he uses for both border and frontier.

[80]"Long live Mexico!" was a phrase which seems to have continued to carry many different levels of meaning within Santiago's later life and his children's lives, and his son Rev. M. F. Tafolla was known to shout "*¡Qué viva México!*" daily at 5 a.m. at the top of his lungs, when immersing himself in his morning bath. Perhaps the irony of surviving such cold water and the gratitude of survival and the warmth of a welcoming home was transmitted in the oral family legends. Stories about M. F. Tafolla, Santiago's son's usage of this phrase were verified by his daughter-in-law Mary D. Tafolla (interview January, 2009) and daughter Esther Tafolla (interview, June,1971.) Tafolla, Mary D. Personal interview conducted by Carmen Tafolla. 18 January 2009. Tafolla, Esther. Personal interview conducted by Carmen Tafolla. 17 June 1971.

[81]Guerrero is a city in the north-eastern Mexican state of Coahuila.

[82]Anastasia's family and the Mercados lived on the Medina at a location very close to the outskirts of San Antonio. The Herreras, Anastasia's sister's family also lived on the Medina, and many of their descendents still do.

[83]Fidel's English translation adds (possibly again from family stories, since Santiago was his grandfather, and the widow of Mercado was his own grandmother) "They say that his desertion from the Confederate Army preyed heavily on his mind, and that in one of these depressive moods he took his life." Santiago's original translates "he lost his sanity, and in a state of dementia . . . "

Other family stories mention an "accident" with a gun, and great psychological pain in the month preceding his death.

[84]The original manuscript says Coley, but Fidel remembered Santiago pronouncing it "Cooley."

[85]In modern days, it is again called Piedras Negras.

[86]The Spanish original says "mis cuñados" (or in Santiago's spelling, "mis cuyados") which could also be translated, "My brothers-in-law" but family lore verifies that it was a family there, so the translation "My brother-in-law and his family" is selected here.

[87]While the original states "agarró frío y murió" (caught cold and died), Fidel translates it as "caught pneumonia at the time she was in labor, and . . . passed away." Family stories provide some support for Fidel's more detailed interpretation.

[88]The children of this marriage were Mollie, Juanita, Catarina, Pete, and Santana. When Juanita Torres de Tafolla died, many of the extended family helped raise the children so Santiago could travel to work. Tafolla, Sarah. Personal interview conducted by Carmen Tafolla. July 18 1964.

[89]Indianola, Texas, now a ghost town, was once one of the most important Texas ports along the Gulf of Mexico. Established in 1846 as Indian Point, it changed its name to Indianola in 1849 after a spurt of growth, in which stagecoach service began and it became the chief port through which European and American immigrants flowed into western Texas. The town soon expanded three miles down the beach to Powderhorn Bayou when Indianola was chosen as the terminus to Charles Morgan's New York-based steamship line. Powderhorn was in the same vicinity.

In 1856 and 1857 two shiploads of camels were offloaded at Indianola. Under the direction of Secretary of War Jefferson Davis, the animals were used in an experiment to transport military supplies through the southwestern United States. A hurricane with 110-mile-an-hour winds blew the town away in 1875 and after an even more destructive hurricane in 1886, the town was not rebuilt. Indianola is located on the Texas Gulf Coast about ten miles from Port Lavaca. *Charles Morgan and the Development of Southern Transportation* by James P. Baughman and published by Vanderbilt University Press, Nashville, 1968.

[90]Mercado, former comrade in arms in the Confederate Army, had deserted the night before Santiago and the remaining *mexicanos* did. After Corporal Juan J. Mercado's death, his widow, Señora Anastacia Salinas de Mercado returned to San Antonio, where her mother and sisters had for many years had a seamstress shop downtown. Anastacia Salinas, born April 15, 1842, told many stories of her courtship by Santiago, including his frequent visits to her house in the month and a half preceding their marriage. See marriage license, June 14, 1869, San Antonio, Texas.

[91]Son, Juan J. Mercado, Jr., and daughter, María Mercado, both became very close to Santiago whom they called Papá. Juan J., Jr, was to later become a Spanish teacher and a preacher, and the translator of *When the Roll Is Called Up Yonder* to Spanish as *Cuando allá se pase lista*, still sung in Methodist, Church of God and Presbyterian churches today.

[92]This very historic crossing on the Medina River is where the Herrera family had long lived, and Anastacia's sister, Sabinita, lived there with her husband, Blas Herrera, Jr., son of the Blas Herrera who had been sent by Juan Seguín to spy on the Mexican Army, and who reported back to Sam Houston that Santa Ana's army was marching on San Antonio, before the battle of the Alamo.

[93]There exists a very popular "Corrido de Jacinto Treviño," many official reports, and much legend in Texas about a Jacinto Treviño who fought against the injustices perpetrated against *mexico-tejanos* by the Anglos, and who ambushed the Texas Rangers in 1911. Said Treviño would have been too young to match this "pariente," but the tradition of naming children after parents and other elder relatives certainly leaves open the possibility of this individual being a father or older relative of the legendary Jacinto Treviño. García, Marina, and Kyle Putnam. "Corridos y Tragedias de la Frontera." 5 Jun. 2001 <http://www.laits.utexas.edu/jaime/jnicolopulos/cwp3/icg/index.html>.

[94]Original says "indianas y imperial", which are two kinds of cotton-based fabrics, the latter heavier and more richly decorated than the first. The New Revised Velásquez Spanish-English dictionary (Chicago, 1974) translates "indiana" as "chintz" while the *Diccionario Castellano enciclopédico*, Novisima edición (Paris, 1927) describes it as "tela de lino o algodón pintada por un solo lado." Imperial, a finer cloth, is also used as "tejadillo o cobertura de las carrozas" Fashion specialist and sales buyer Herlinda Marroquin remembers having dresses made of both, but states that the "imperial" was much heavier, and warmer, and "too hot for summer usage." *The New Revised Velasquez Spanish-English Dictionary*. Chicago, 1974. *Diccionario Castellano Enciclopédico*, Novísima Edición. Paris, 1927. Marroquin B. Guerra, Herlinda. Personal interview conducted by Carmen Tafolla. April 3 2009..

[95]The Spanish original says *pesos*.

[96]Page 80 of Commissioners' Minutes Book "B" of the County Commissioners' Court of Bandera County, May 29, 1876, states: Charles Montague, Clerk, $2,000.00 . . . James Tafolla, as Justice of the Peace Precinct 2, bond in the sum of $500.00 . . . Octaviano Herrera as Constable Precinct 2 $500.00, F. L. Hicks, Inspector of Hides and Animals, $1,000.00. And it appearing to the Court that said officers had each filed his said bonds within the time prescribed by law and that the sureties upon each said bonds were good and solvent men and citizens of Bandera County, it is therefore ordered that the aforesaid bonds be and each of them is hereby approved and ordered of record . . .

LAS MEMORIAS DE UN MEXICOAMERICANO EN LA CONFEDERACIÓN

PEARSALL, TEXAS
16 DE SEPTIEMBRE, 1908

Como varios de mis hermanos en la fé me han animado a que escriba mi biografía y como también mis hijos lo desean, hoy me atrevo a comenzar a hacerlo, aunque me considero incapaz para escribir con propiedad para presentar ante el público debidamente mi escritura, pues sé muy bien que carezco de educación suficiente, pero confiando en la indulgencia de mis amigos, y con el fin de obsequiar sus deseos, me propongo darles mi biografía aunque imperfectamente, pero lo mejor que me sea posible esperando que los que lean estas mal escritas líneas, se fijen que todo lo que escribo es la verdad de lo que me ha pasado y lo escribo sin parcialidad ninguna, no buscando mi honra, sino exponiendo todas las circunstancias que en la Providencia de Dios he experimentado.

Comienzo mi biografía hoy a la edad de 71 años, y 23 días. Doy gracias a Dios que habiendo pasado por tantos peligros me permite dar testimonio de las muchas bendiciones que por Su Gracia Divina me ha concedido.

Cuando yo vine a Texas, no había otro Tafolla en todo el Estado, y ahora cuento con siete hijos, cinco hijas y 45 nietos.

—Santiago Tafolla

CAPÍTULO 1

Nací en Santa Fe, Nuevo México, el día 23 de julio de 1837. Mi padre se llamaba Mariano Tafolla, y mi madre Josefa Gutiérrez, ambos nativos de Santa Fe. Mi padre era militar y oficial en el ejército de México, y viniendo en camino de la capital de México encargado de un tren de carros en el que traían una grande cantidad de dinero, una noche asaltaron los indios al tren que estaba durmiendo en un lugar que le llaman Galisteo, como unas 30 o 40 millas de Santa Fe. Nomás mi padre quedó muerto en el campo. La escolta que acompañaba al tren que serían como 40 hombres, todos huyeron y vinieron a dar las nuevas a Santa Fe.

Las tropas que salieron para el lugar hallaron a mi padre muerto con el corazón atravesado por una flecha. Se decía que los mismos soldados que formaban la escolta habían matado a mi padre por robarse el dinero. En eso se les formó juicio, pero yo estaba muy joven para poder acordarme, pues todo esto lo supe por tradición de mi madre y mis hermanos. Yo apenas tendría tres años cuando mi padre murió.

Cuando yo tenía siete años de edad murió mi madre, y mi hermano mayor recogió a toda la familia que había sin casarse, que fue una hermana mía, ya señorita grande y a mí, que era el menor de toda la familia. Mi hermano mayor se llamaba Lorenzo. También había otro de los [hermanos] soldados que se llamaba Rafael, que se le seguía a mi hermano Lorenzo. Mis hermanas mujeres [eran] cinco, cuatro ya casadas y una soltera, y se llamaban [las casadas] Rosario, Cruz, Acención [y] Refugio. La señorita se llamaba Dolores.

CAPÍTULO 2

En aquellos tiempos el pueblo do Santa Fe se llamaba La Villa, y los de los pueblos circunvecinos les decían "los villeros" a los de Santa Fe. La familia Tafolla vivía en un lugar que le decían El Realito. Allí tenían sus propiedades de terrenos. Las labores eran de riego. Mi hermano sembraba varias legumbres, pero principalmente maíz y trigo. Siempre tenía tres o cuatro peones que trabajaban en la labor, y a mí me hacía trabajar al parejo de ellos. También tenía cuatro o cinco burros, y en tiempo de frío mandaba a los peones al monte a traer leña en los burros y me hacía que yo fuera con ellos.

Salíamos de la casa como a las dos o las tres de la mañana con unos fríos terribles, con la tierra toda cubierta de nieve, pues en esa parte de Nuevo México las nieves cubren la tierra desde el mes de noviembre hasta principios de abril. El trabajo que hacía yo con los burricos (pues así les llamaban) era juntar la leña que ellos quebraban y hatajar los burros, y demás cosas que me mandaran y esto lo tenía que hacer a la carrera y aunque yo andaba helándome de frío. Si se enojaban conmigo, me arrebataban a patadas, y de cada patada me iba resbalando por sobre la nieve como diez pies y quedaba tan adolorido que parecía que todos mis huesos estaban quebrados, y ni a quien quejarme porque mi hermano era lo mismo de cruel conmigo.

Cuando volvíamos del monte, sería a las 10 de la mañana, la leña valía mucho y se vendía hasta a peso la carga y eso era quehacer mío. Los peones se iban a sus casas y yo iba a vender la leña antes de ir a la casa. A veces cuando volvía a la casa, era mediodía y a esas horas me desayunaba.

Durante el resto del día yo era el de los mandados y si me tardaba en un mandado, mi hermano era tan cruel que me amarraba y me quitaba la camisa y con un cuarta de azotar caballos, me azotaba las espaldas hasta que me corría la sangre, y por mucho tiempo me quedaban los verdugones, de tal manera que no podía acostarme boca arriba.

Después de recibir algún de estos castigos, mi hermanita que vivía también en la misma casa, me agarraba allá a solas y lloraba conmigo, pero yo le decía: "No llores, manita, alcabo nomás que yo esté más grandecito me huyo de aquí". Y esto la hacía llorar más. Tendría yo diez años cuando ella se casó. Entonces, yo me quedé sin quien simpatizara conmigo. La esposa de mi hermano no se compadecía mucho de mí. Ellos tenían una hija que se llamaba Francisca, y como era la única hija, la querían mucho y a veces sufría yo por causa de ella.

CAPÍTULO 3

Mi hermano era un hombre respetable y, aunque no era rico, tenía buena comodidad. También había sido oficial en el ejército mexicano. Era lo que se llama en el Romanismo "un hombre religioso". Muy devoto de la Virgen de Guadalupe, todos los años, en el aniversario de esa virgen, que es el doce de diciembre, nueve días antes de esa fecha comenzaba a rezarse el rosario y la novena de esa virgen todas las noches en su casa. Convidaban mucha gente, se llenaba la casa. Y después del rosario y la novena, se servía una cena a todos los convidados, y el día 12 en la noche, velaban y rezaban hasta media noche. Y entonces se concluía con un baile hasta que amanecía.

Tenía la costumbre mi hermano de hacer caridad el día 12 de diciembre, a dos ancianos que él escogía de los pobres del pueblo. Les cargaba de maíz y frijol y de todo lo que tenía y también de piezas de manta, y de imperial y otras ropas.

En esto era muy religioso, pero también era jugador y a veces no venía a la casa en toda la noche por estar en la jugada. Y cuando ganaba, venía muy contento, pero cuando perdía, venía borracho y echando alaridos. Y nomás que oíamos el alarido, ya no hallábamos en dónde meternos toda la familia, porque venía muy enojado. Y siempre se encomendaba a la virgen de Guadalupe para ganar. A veces cuando perdía, se emborrachaba de sentimiento, y venía a media noche golpeando la puerta, gritando y echando maldiciones. Y nomás lo oíamos y ya estábamos temblando de miedo. Y en tales ocasiones, no se acordaba de religión, ni de la Virgen de Guadalupe.

Y bajo tales influencias me crié yo hasta la edad de once años.

CAPÍTULO 4

Por el año 1848, en el mes de septiembre, nos juntamos yo y un muchacho, hijo de una hermana de mi madre, que vivía con unos parientes y pasaba una vida semejante a la mía y nos pusimos de acuerdo para huirnos juntos y ir a Las Vegas, un pueblo como a setenta o ochenta millas de Santa Fe.

El día que salimos anduvimos unas veinte millas. Durmimos en el campo sin tener con qué taparnos, ni qué comer.

En la mañana seguimos el camino, y cosa de las diez, llegamos a un lugar donde había dormido un tren de carros de americanos. Todavía estaban las lumbres ardiendo. Allí nos hallamos pedazos de pan y de jamón, y con eso nos desayunamos. Y seguimos nuestro camino.

A cosa de cinco o seis millas, agarramos una vereda que atravesaba unos arroyos muy hondos por donde caminaba la gente de a pie y de a caballo, mientras el camino de los carros daba una vuelta de cosa de tres millas. Al salir al lugar donde se juntaba la vereda al camino, nos sorprendimos al encontrar el tren que venía delante de nosotros. Eran como cincuenta carros de bueyes, y en cada carro iban tres o cuatro hombres como pasajeros, en algunos más. Este tren era de varios dueños, pero uno solo le hacía de comandante. Habían ido de Independencia, Estado de Missouri, hasta Santa Fe, con flete del gobierno para las tropas que había en Nuevo México, atravesando los llanos estacados que en ese tiempo era un desierto lleno de cíbolos que se veían de a miles, cubiertos los llaños con ellos. También había muchos venados y berrendos y caballada mesteña. Y también había muchos indios broncos que estaban de guerra con todo el mundo civilizado, y esto hacía muy peligroso caminar en esos tiempos por esos desiertos. Tanto que el tren de carros caminaba por dos caminos, quedando una distancia como de diez varas el uno del otro. Se dividía el tren, y la mitad de los carros iba por un camino y la otra por el otro. Y en caso de ataque de indios, los carros de adelante se juntaban y los demás carros se arrimaban uno tras de otro. Y así formaban una muralla o fortín. Y de esta manera se formaba un corral todas las noches.

CAPÍTULO 5

Cuando yo y mi primo hermano (que se llamaba Plácido) vimos los carros, nos quedamos parados de susto, como que no hallábamos qué hacer hasta que un hombre de los que iban en un carro nos dijo en muy mal español, "Machacho, ¿quiero vamos Estados Unidos?" Yo le contesté, "Sí, yo quiero," y le dije a Plácido, "Vente, vamos" pero él se quedó parado porque les tenía miedo a los americanos, mientras yo fui y me subí al carro. Lo llamaba yo con la mano, hasta que lo perdí de vista y esa fue la última vez que vi a mi primo hermano.

El hombre que me invitó que subiera al carro se llamaba Robert A. Matthews y desde ese día se hizo cargo de mí. Tardamos como dos meses mas o menos atravesando lo que se llaman los llanos estacados o Staked Plains.

Durante esta travesía no dejé de ver algunas cosas curiosas y experimentar algunas aventuras. En esos llanos vi mucho terreno donde habitaban muchas tusas que en inglés les dicen "prairie dogs," y los lugares más poblados por ellos les dicen "dog towns".

Llevábamos en el tren doce indios Shahuanos que en inglés creo les dicen Shawnee Indians. Estos eran empleados por los dueños del tren como cazadores y exploradores que siempre iban detrás y delante del tren como espías. Llevaban carabinas de pelo, y eran muy buenos tiradores. Todos los días mataban venados y berrendos y también cíbolos. Las tusas o "prairie dogs" salían todos los días de sus cuevas y se paraban en las patitas de atrás y le ladraban al tren por un lado y otro del camino.

Un día uno de los indios le tiró a uno de ellos y lo hirió, echándole fuera las tripas. Yo por curiosidad lo llevé al campo donde dormimos esa noche, y por estar jugando con él me agarró un dedo con los dientes, y yo no podía hacer que me soltara hasta que los compañeros me lo quitaron. Sufrí por algún tiempo de la mordida. Se me hinchó la mano y el brazo. Después supe que eran ponzoñosos. Vivían con las víboras chilladoras en las cuevas.

El tren no caminaba los domingos y un domingo fueron unos americanos a bañarse. Yo fui con ellos, pero no me bañé yo con ellos porque tenía vergüenza desnudarme delante de ellos. El agua estaba bastante honda para nadar y había mucho zacate tule del que crece en el agua y yo no creí que estaba tan hondo. Cuando ellos se fueron, me fui a bañar y me fui hasta donde estaba muy hondo, y como no sabía nadar, creí que me iba a hogar debajo del agua. Se me ocurrieron miles de pensamientos en un momento. Creí que me iba a hogar, y que era castigo de Dios por habérmele huído a mi hermano y muchas otras cosas se me vinieron a la mente en un momento. Pero no perdí mis sentidos. Me agarré del zacate y me fui al plan y teniéndome del zacate anduve por debajo del agua hasta que me paré y saqué la cabeza. Nunca se me ha olvidado este acontecimiento.

Como a un mes quince días o dos meses llegamos a las primeras habitaciones americanas en la frontera de Kansas. De allí tomamos la diligencia o sea "the Stage[coach]". El primer pueblo americano que ví se llamaba Council Bluff.[1] Estaba situado a orillas de un río. No sé cómo se llama [el río] porque, como no sabía el inglés, no sabía los nombres de los lugares y también estaba muy joven para interesarme en esas cosas. Llegamos como a las once de la mañana.

Durante el tiempo que estábamos esperando la salida en el hotel pasó otro acontecimiento que nunca se me ha olvidado. El cochero que arreaba la diligencia llegó y se paró en la puerta del hotel. Se apeó y entró a la sala de espera y sacando una guardamoneda[s] lleno de dinero, comenzó a insultar, según parecía, a todos los que estaban presentes, que serían como quince o veinte persones. Parecía que [quería] desafiar a todos porque daba vuelta por toda la sala con la portamonedas en la mano. Se la sacudía en la cara [de todos] y no había quién le dijera nada.

Había un hombre sentado en una silla que, al parecer, estaba enfermo porque estaba muy pálido. A éste se le acercó el cochero, y pegándole la portamoneda[s] a la cara, le hablaba al parecer con mucho furor. Y el otro se ponía más pálido. Cerca del enfermo esta[ba] sentado un señor anciano alto, delgado que también parecía estar enfermo. Tenía en la mano un bastón grueso y muy macizo. En una vuelta que el cochero dio para afuera, donde estaba el coche, este señor anciano lo siguió, y al volver [el cochero] a querer entrar [por] la puerta, el anciano le pegó un palo entre medio de los dos ojos y el cochero cayó para atrás en el empedrado y al irse a levantar, le pegó otro palo en la cabeza y [el cochero] entonces cayó en la calle detrás del coche, y el anciano se fué sobre él y le pegó muchos palos en las espaldas y el cochero nomás gritaba y toda la gente que estaba presente decía, "Hit him, hit him again". A lo último, el anciano lo dejó, sin hablar [ni] una palabra desde que comenzó hasta que acabó. Entonces el cochero se subió al coche y tomó las riendas y dos señoras que ya estaban en el coche comenzaron a gritar asustadas. Algunos hombres agarraron los caballos y otros apearon a las señoras y el cochero se fue gritando.

En cosa de quince minutos, volvió con otro compañero y subieron todos los pasajeros, yo entre ellos. Yo no sabia qué era lo que pasaba, nomás lo que veía, pero lo que oía no lo entendía. Esa noche llegamos a Independencia, Missouri, un pueblo muy bonito.

Esa noche en el hotel encontré [a] un mexicano y me preguntó cómo me gustaba el pueblo. Le contesté que me gustaba mucho y me contestó, "No se encante, jovencito, esto [no] es nada en comparación con México".

CAPÍTULO 6

El día siguiente salimos de Independencia. Íbamos en el coche yo y Mr. Matthews y otros tres hombres y una señorita. Al medio día cambiamos [de] diligencia en un pueblo cuyo nombre no me acuerdo. Esa noche llegamos a unos cerros muy altos y muy peligrosos, pues el camino quedaba muy parado y los caballos, que eran cuatro muy gordos y briosos, empezaron a cortarse antes de llegar a media subida. Entonces [los pasajeros] determinaron que se apearan los hombres y yo y la señorita nos quedamos en el coche. Los hombres se fueron adelante creyendo que el coche los alcanzaría después de subir la cuesta.

Yo y la señorita nos quedamos en el coche, yo con un fúsil de dos tiros del Mr. Matthews en las maños. Los caballos dieron unos cuantos pasos y ya no quisieron estirar. Se cortaron y empezaron a reparar. Y viendo el peligro de que se volcara el coche, vino uno de los cocheros (porque eran dos) y nos dijo que nos apeáramos y nos fuéramos con la gente que iba delante. Yo con el fúsil en la mano, nos fuimos por el camino con la esperanza de alcanzar a la gente, o que nos alcanzara el coche. Sería[n] como las doce de la noche, y andaríamos como una milla cuando llegamos a la cumbre de la cuesta.

Ni yo ni la señorita habíamos hablado ni una palabra. De repente se paró ella y me dijo, "Will you take care of me?" Yo, no entendiendo lo que me decía, le contesté "No, sir." Al momento se paró contra un encino que estaba cerca de ella. Yo también me paré como [a] diez yardas de ella pensando qué hacer. Comprendí que la señorita estab[a] asustada y que yo le había contestado mal, pero ni modo de explicarme, pues ni ella me entendía a mí, ni yo a ella.

Estaríamos así como quince minutos cuando vino uno de los cocheros a caballo, y me dió a entender que algo había sucedido y que fuera a alcanzar a la gente y les avisara. Corrí tan veloz como pude en la oscuridad de la noche y, como a media milla, hallé a los compañeros parados, esperando el coche. Como yo pude, les dí a entender que algo malo había sucedido. El padre de la señorita se asustó, y me preguntaba si algo le había sucedido a la señorita. No pudiéndome yo explicar, Mr. Matthews, que hablaba un poquito [de] español me preguntó, "¿Está mujer malo?" Yo le dije, "No, señor, nada le ha sucedido a la señorita".

Entonces el padre se calmó, y todos juntos nos volvimos hasta donde estaba el coche que se había volcado, y había rodado como cien varas por la ladera y se había hecho pedazos. Allí nos estuvimos hasta que amaneció, calentándonos, y platicando los americanos y yo oyendo, pero sin entender nada. Como a las ocho de la mañana, vinieron con un carro, y montando todos en el, continuamos nuestro camino.

En ese tiempo, esa parte del estado de Missouri estaba muy poco habitada. En el primer pueblecito que llegamos nos esperaba la diligencia con cuatro caballos. De allí montaron dos señoras. La señorita que mencioné arriba y su padre se quedaron. Durante el día llegamos a un lugar donde había mucha arena y los hombres que venían en el coche se apearon como lo hacían siempre que el camino estaba pesado. Debo decir que dos hombres eran compañeros del Mr. Matthews y venían con él desde Nuevo México. Uno se llamaba Love y el otro McGoffin. Ellos nos acompañaron todo el viaje hasta que llegamos a Washington.

CAPÍTULO 7

Las señoras que venían en el coche venían riéndose y platicando algo con respecto a mí, porque me miraban y yo comprendí que me querían hablar. Y venía pensando "[Si] me hablan qué les diré?" Dicho y hecho, una de ellas se arrojó y me dijo: "Do you speak English?" Yo le contesté, "No, sir." A mi contestación tan brusca y fuera de orden las dos soltaron la risa y yo me quedé un poco avergonzado. Y observándome, una de ellas me dijo: "That *is* English." Y en verdad era todo el inglés que yo sabía.

Para no alargar más esta parte de mi historia, diré [que] caminamos por diligencia como diez días hasta que llegamos a San Louis, Missouri. No sé en qué fecha llegamos. Esa noche dormimos en un hotel, y estuvimos allí dos días. El [primer] día que amanecimos allí, como a las diez de la mañana, me salí a dar una vuelta en la calle. Había andado como tres cuadras cuando oí [el] silbido de un barco de vapor que iba llegando al muelle. Yo, no habiendo oído nunca cosa semejante, partí huyendo toda la calle rumbo a donde [se] oía el ruido.

Cuando llegué al río vi la línea de embarcación por un lado y otro del río, y me pareció la cosa más maravillosa que había visto en el mundo. Y en verdad que lo era, pues nunca había visto ni siquiera una esquifa. Por más de dos horas me divertí viendo las embarcaciones. Había como cincuenta barcos de vapor grandes y muchas escunas[2] y barquitos de vela. Cuando quise volverme para el hotel ya no sabía qué rumbo tomar, y como no sabía cómo se llamaba el hotel, ni entendía el inglés, no podía preguntar. Anduve así perdido hasta como a las tres de la tarde.

Llegué a una barbería donde estaba un negro cuarterón que parecía mexicano y le hablé, pero él no me entendió. Pero me agarró de la mano y me llevó a una casa donde estaba una mujer mexicana y le dijo que me dijera que si yo quería vivir con él, él me enseñaría el oficio de barbero. Yo le expliqué a la mexicana cómo y de qué manera había ido yo a St. Louis, y ella le dijo al negro que se fuera y me dejara allí, que ella me conduciría a mi lugar y ella me aconsejó que no me juntara con los negros y me mandó al hotel con un muchacho americano. Mr. Matthews tenía mucho cuidado de mí y me había andado buscando.

De San Louis tomamos un "steamboat" hasta Cincinnati, Ohio y de allí a New York City y de allí a Baltimore. Yo me acuerdo que tocamos todos estos lugares y también Buffalo, New York, pero no sé cuáles serían sus negocios que traían en mano. El caso es que los tres compañeros que he mencionado llegaron juntos a Washington. Mr. Love tenía un hermano allí que era delegado del congreso, no sé de qué estado.

Paramos en The United States Hotel. Mr. Matthews arregló con el dueño del hotel que se llamaba Hackney de que me dieran la comida y cama por el servicio que yo pudiera dar. No me acuerdo en qué mes fue, pero creo que fue en diciembre cuando llegamos a Washington. Estuve en ese hotel como diez meses. En este tiempo aprendí a hablar en inglés bastante para desempeñarme en todo lo que tenía qué hacer. Y llegué a ser el favorito del dueño del hotel y los demás de su familia. Yo hacía mandados a todos los hospedados. Me gané también la simpatía de ellos. Y mi quehacer era de "messenger", contestar al llamado de todos los cuartos cada vez que sonaban la campanilla. Ya sabía todos los números de los cuartos y me ganaba algún dinero con los mandados que hacía.

En ese tiempo vinieron unos señores mexicanos de la capital de México. Yo creo [que] traían negocios del gobierno, o serían muy ricos, porque gastaban mucho dinero. Me ocupaban para que les interpretara en algunas cosas, aunque traían un intérprete. Estos señores, don Antonio Arrasol y un hijo de él, que también se llamaba Antonio, y otro señor, que se apellidaba Balgán. (No recuerdo su primer nombre.) El intérprete se llamaba Juan Arrasol. Era hombre ya mayor y de muy buena educación. Estos señores se distinguían de todos los demás abonados. Mandaban ponerse mesa separadamente y les servían manjares muy escogidos y vinos finos a todas horas. Y eran muy liberales con su dinero. A veces me daban cinco pesos para que les trajera algún mandado. Y si sobraba uno o dos pesos me los daban. También llegué a ser favorito con ellos.

Yo no molestaba a Mr. Matthews por vestido porque yo ganaba bastante con mis grangeos para vestirme. Durante este tiempo, se salió del hotel el velador que velaba de noche la oficina, y me pusieron a mí en este quehacer. Y en esto también tenía algunas ganancias de los pasajeros que venían y salían de noche. Durante el tiempo que estuve de velador me quitaron el trabajo de mensajero y tenía todo el día libre para dormir. Y en este tiempo aproveché la oportunidad de ir a la escuela pública.

Durante este tiempo hubo la elección del presidente de [los] Estados Unidos y fue electo el presidente Pierce que corrió contra el General Scott, que fue el General en Jefe durante la guerra entre México y los Estados Unidos de 1846 y 1847.

CAPÍTULO 8

Como a los diez o once meses de estar viviendo en Washington y en el United States Hotel, no sé por qué causa Mr. Matthews se disgustó con el Mr. Hackney, el dueño del hotel, y cambió de residencia. Él estaba empleado en la capital en asuntos del congreso. Nunca supe yo qué sería su empleo. Cuando se cambió, rentó un cuarto en una casa de hospedaje de una señora que se llamaba Mrs. Mary A. Davis. Tenía como seis o siete cuartos amueblados y los rentaba. Estaba frente al National Hotel, en Pennsylvania Avenue, la misma calle donde estaba The United States Hotel. Algunos miembros del congreso se hospedaban en esa casa.

Mr. Matthews me sacó del lugar donde yo estaba y me trajo a la casa de Mrs. Davis y hizo arreglo con ella de que yo le sirviera de mandadero y demás quehacer que pudiera desempeñar en la casa, con la condición de que fuera a la escuela durante las horas de escuela. La señora era muy buena señora. Vivía su madre con ella, que era muy anciana, y tenía un niño de seis años y una niña de ocho. Era toda la familia de la casa. Los cuartos los rentaba los más baratos por un peso al día y algunos por dos pesos cincuenta centavos. Los cuartos eran muy caros en Washington en ese tiempo.

La señora Davis me recomendó a una escuela católica que estaba cerca de Georgetown, aunque yo no creo que ella fuera miembro de ninguna iglesia, porque nunca la vi que fuera a ninguna, ni tener oficios religiosos en la casa. Sin embargo, tenía a su niña en una escuela católica y tal vez creyendo que, como yo era mexicano y católico de nacencia, me correspondía ir a esa escuela.

Todos los maestros que yo conocí [eran] los que yo conocía en mi tierra y eran legos o [sic] majoros.[3] Era una escuela de 3 o 4 departamentos. A mí, me tocó a uno donde estaban los niños mas pequeños. Desde el primer día observé que el modo de enseñar era muy diferente al de la escuela pública donde había estado. Lo más del tiempo lo gastaban en rezar. Lo primero que se hacía en la mañana era ir a la misa que se celebraba en un capilla que estaba a un lado del mismo edificio.

Nunca he visto hombres tan crueles con los niños como esos hombres. En el departamento donde yo estaba había niños que no pasaban de seis años y les daban unos castigos tan crueles que me hacían recordar de lo que yo sufría en mi infancia, cuando en poder de mi hermano Lorenzo. Por la más leve cosa, castigaban a los niños de seis o siete años. A veces los agarraban de las orejas y los levantaban como tres pies y los aventaban y no habían de llorar. Yo era de los más grandes en ese departamento y me determiné que el día que me dieran un castigo de esa manera, dejaría la escuela.

La señora Davis pagaba 50 centavos por mes por mí. Sufrí algunos castigos de los que no eran tan severos porque parecía que me veían con alguna consideración, y me creía que tal vez sería por ser yo mexicano. Pero un día, no sé porque, el lego me llamó de mi asiento. Algo hize, no me acuerdo qué, pero él me agarró de las orejas (que parece que era su castigo favorito) y no me levantó porque [yo] ya estaba pesado. Pero me dejó las orejas coloradas.

En la tarde le dije a la señora Davis que ya no quería ir a esa escuela y le platiqué de lo que allí pasaba. Y otro día en la mañana, me dió 50 centavos y le escribió una notita al lego diciéndole que yo no quería ir mas a la escuela. Todo lo que aprendí en cosa de tres meses que estuve allí fué a rezar la Salve y la Ave María y el Padre Nuestro y el credo, en inglés, todos —pero de lectura, nada.

De allí, volví a la escuela pública donde estaba antes. Estuve en la casa de la señora Davis como un año. Durante este tiempo conocí al ministro plenipotenciario de México. Los domingos me daban libros luego que se acababa mi quehacer de la mañana y yo logré visitar al ministro que me recibía con cariño. Tenía algunos jóvenes mexicanos de la capital de México y estaban yendo al colegio, y ellos me trataban muy bien y travesiaban conmigo. Y me decían el nuevo mexicano, porque era de Nuevo México.

También visité al ministro de España. Fui a su casa, y me recibió bien. Me hizo varias preguntas y me preguntó si mi visita era con el fin de que me diera algo. Y le dí las gracias y me despedí de él. Se llamaba el señor Calderón. El ministro mexicano me parece que se llamaba Matías Barrientos.

CAPÍTULO 9

Durante este tiempo Mr. Matthews se casó y determinó ir con su esposa a hacer una visita a su tierra natal que es Talbaton,[4] estado de Georgia. Me llevó consigo hasta Charleston, South Carolina, y allí me dejó con un hermano de él que era albañil. Estuve allí como tres meses, y su hermano me mandó para Georgia a casa de otro hermano que vivía en Muscogee County. Se llamaba Charles Matthews. Era doctor. Vivía en una plantación que le pertenecía. Tenía algunos esclavos. Su esposa tenía una escuela de las que se llamaban "County Schools". Allí fui a la escuela cosa de tres meses.

Durante las horas y los días que no había escuela, trabajaba en lo que me mandaban. Pero un día hubo un disgusto en la escuela y los muchachos me culparon a mí de haber sido el causante del disgusto. Y se lo dijeron al doctor y éste vino y me regañó como queriéndome asustar, y yo le dije que él no me asustaba, pues yo sólo le tenía miedo a uno. Y él me preguntó quién era ése y yo le dije: "Dios". Entonces él se sonrió y me dejó solo en la casa de la escuela que estaba como a cincuenta varas de su casa de residencia.

Yo me sentí ofendido, y me quedé pensando qué debía hacer. Tenía vergüenza ir a la casa, y mejor me determiné a irme de allí. Sería como a las ocho de la noche. El camino de fierro que se llamaba The Muscogee Railroad pasaba cerca del lugar, y yo había oído decir que había como veinticinco o treinta millas a Columbus, y me fui por todo el camino de fierro. Caminé toda esa noche y otro día todo el día, sin comer más que duraznos que había en alguna[s] hortalezas en la orilla del camino.

En la noche llegué a Columbus como a las ocho. Me fui por una calle hasta que hallé un hotel y me senté en unos bancos que estaban al frente. Como había mucha gente allí y que pasaban y entraban, y salían del hotel, nadie se fijó en mí. Allí me quedé dormido, como que venía muy cansado. Otro día me determiné a buscar trabajo. Fui a una fábrica donde trabajaban la lana y me dieron un trabajo de subir lana de la que lavaban en el piso de abajo hasta el tercer piso, por una escalera. Y la tenía que subir en una canasta muy pesada. Subí tres viajes y me faltó el aliento, pues hacía dos días que no comía. Fui y le dije al de la oficina que no podía hacer el trabajo porque era muy pesado para mí y que no había comido por dos días. Y me dijo que no tenía otro trabajo, pero que fuera y buscara en dónde me dieran la comida, que había varias fonditas en donde comían los trabajadores de la fábrica. Que les dijera que yo estaba trabajando.

Me fui muy desconsolado, considerando de que aunque hallara la asistencia, el trabajo estaba muy pesado para mí. Mr. Matthews estaba todo este tiempo en Talbaton y yo me determiné a irme para ese lugar. Pregunté por dónde salía un camino para Talbaton y me fui. Mi esperanza era de hallar duraznos por el camino o alguna otra fruta para comer. Y así fue. A poco de haber salido, como a una milla, hallé una hortaleza con muchos duraznos. Seguí preguntando por el camino a todos los que encontraba. En la noche dormí en el campo. Otro día en la noche llegué a Talbaton, ya noche como a las ocho.

Pregunté por la casa del doctor F. T. Matthews y un muchacho me llevó. Ya allí tenían la noticia de que me había huído, pero no sabían ni para dónde. Mr. R. A. Matthews no estaba en la casa. Él y su esposa estaban en casa de una hermana de él, cuyo esposo era abogado. Se llamaba Cavend[er]. En la casa de Dr. F. T. Matthews me recibieron muy bien, compadeciéndose de mí por los trabajos que había pasado. El día que amanecí allí, vinieron Mr. [R. A.] Matthews y su esposa y también simpatizaron conmigo.

Luego que las gentes supieron que un muchacho mexicano estaba en el pueblo, venían a verme por curiosidad y todos hablaban *of the Mexican boy.*[5] El doctor F. T. Matthews era un hombre muy cristiano y miembro de la Iglesia Metodista Episcopal del Sur. Era jefe de clase y era un hombre muy caballero en todo respeto. Las gentes que venían a verme decían, "He is nearly White," y otros decían, "He is as White as anybody" y el doctor les decía, "Of course he is White." Como en esos tiempos había esclavitud, los blancos no se juntaban con los negros de ninguna manera, y era una degradación para un blanco aún simpatizar con los negros. Me acuerdo que una niña de la señora Matthews le dijo un día a su mamá, "Mama, is Mexican Jim sure enough White?" Y ella le dijo, "Hija, James' blood is as pure from Negro blood as yours or mine."

CAPÍTULO 10

En la casa del doctor, me dieron un lugar como si fuera de la familia. Todos los domingos iba a la escuela dominical con la familia y a todos los demás cultos que había en la iglesia. El doctor acostumbraba tener cultos familiares, a tarde y mañana. Se juntaba la familia en la sala. Él leía la Biblia y hacía oración. En esa casa comencé a leer el nuevo testamento en inglés.

En ese lugar había una escuela que se llamaba The Talbaton Institute, y por influencia del Dr. Matthews fui admitido en ella. El presidente del instituto se llamaba McQueen. La escuela estaba cosa de media milla del pueblo. En esa misma escuela estuvo nuestro obispo Key, pero [algún] tiempo antes de que yo fuera allí. Él mismo me lo platicó. Él conoció a toda la familia Matthews. En ese tiempo había un ministro encargado de ese circuito, que se llamaba Hinton. No me acuerdo [de] sus iniciales pero en una reunión campestre que hubo me convertí. Me acuerdo muy bien que en su oración le pidió a Dios por la conversión del muchacho mexicano que se halla entre nosotros. Y yo nunca he dudado de que Dios contestó esa oración, y espero un día ir a darle la mano en el cielo como se la di ese día cuando hizo el llamamiento y yo creí en mi Salvador con todo mi corazón.

El Dr. F. T. Matthews tenía plantación a cosa de cinco millas del pueblo y tenía como quince negros en ella. Ese año se le salió el mayordomo que era un hombre blanco. (Todos los dueños de plantaciones que tenían esclavos, tenían mayordomos blancos) y como en ese tiempo eran las vacaciones de la escuela, estaba yo en la casa. Y el doctor me puso de mayordomo en la plantación. Había un negro anciano que era el que se entendía con la dirección del trabajo, de manera que mi trabajo era nomás cuidar de darles las provisiones cada semana y guardar el orden, porque se hacía muy necesario que hubiera alguna persona blanca en el lugar a quien se diera cuenta de lo que sucedía. Estuve allí como tres meses.

Durante el tiempo que estuve allí, todos los días salía con un fúsil de dos tiros a dar vuelta alrededor de la cerca de la labor, que sería como dos millas de circumferencia y les tiraba a las ardillas y mataba varias todos los días, y la negra cocinera me las guisaba para comer. Un día que iba yo rodeando la labor, como de costumbre, me dieron ganas de brincar para adentro y puse el fúsil recargado contra la cerca. Y después que brinqué al otro lado agarré el fúsil del caño y lo estiré. El martillo se [en]ganchó en una rama y se disparó el fúsil. Se me resbaló [de] la mano, y lo solté. Me quemó el guante y me voló el sombrero. Y le hizo pedazos la falda de adelante [al sombrero], y algunos granos de pólvora se me enterraron en los labios. También la mano se me chamuscó un poco. Di gracias a Dios que no me maté ese día. Y me puse a pensar, si me hubiera matado, quién sabe cuándo me hubieran hallado. Pero Dios no lo permitió porque tenía un trabajo para mí. Yo siempre lo he pensado así.

Ese año después que vine de la plantación, me vino la idea de comenzar a aprender un oficio. Y así le dije al Dr. Matthews. Mr. R. A. Matthews ya se había venido para Washington. El doctor le habló a un señor que tenía un taller de zapatería con respeto a mí. Este señor se apellidaba Brown. Era un señor anciano muy respetable. Convino tomarme como aprendiz, dándome la comida, y cama y ropa por el primer año. Él no tenía familia, se asistía en un hotel y dormía en la casa donde tenía el taller. Allí tenía un cuarto que usaba como recámara y allí puso una cama para mí. Este señor, como he dicho, era muy respetable. Le decían el capitán Brown, porque había sido capitán del ejército americano durante la guerra de Estados Unidos y México. Fue hasta la capital de México. Era miembro de la iglesia Metodista y muy puntual asistía a los cultos el domingo y [entre] la semana. Me trataba muy bien, pero tenía sus ratos de genio duro.

Haría como tres meses que estaba trabajando con él cuando un sábado en la tarde, después que cenamos en el hotel, de allí mismo me llevó a una tienda y me compró un vestido a gusto de él y un sombrero del mismo estilo que él usaba, que era de esos altos, de pelo, que les decían los muchachos americanos "two-story hats" (sombreros de dos pisos.) Y me quise oponer a que me comprara sombrero de esa clase, pero él me dijo, "¿Eres tú mejor que yo?" Ya yo sabía que lo que él decía era ley. Otro día me mandó que me vistiera para ir a la iglesia. Yo siempre iba con él. Me vestí, y como yo no quería usar el sombrero alto, me puse el viejo. Pero él me hizo que me pusiera el nuevo y me regañó, diciéndome que no era yo mejor que él.

Yo tenía tanta vergüenza pa' que me vieran los muchachos, que nomás veía un muchacho y me quitaba el sombrero. Me parecía que habían de decir, "Look at Mexican Jim's hat." Pues todos me conocían por "Mexican Jim." Mr. Brown notó que yo tenía vergüenza usar el sombrero y cuando vinimos de la iglesia, me regañó. En la noche que fuimos a la iglesia me puse el sombrero viejo. Y él no me dijo nada. Otro día en la mañana, me levanté a poner lumbre y preparar el cerote para el uso de los oficiales como era mi costumbre, y él no me dejó. Me dijo enojado: "Ya no te necesito más. Vete para tu casa y dile al doctor que yo hablaré con él".

Yo me fui callado, ya sabía que no me valían palabras. El doctor habló con él, y él le dijo que ya no me quería porque yo no podría nunca ser zapatero, porque era muy orgulloso.

En ese tiempo había un alemán que se llamaba Meyers que tenía una zapatería y él me ocupó para que yo le ayudara en el trabajo que yo podía hacer. Estuve ayudándole como dos meses. En ese tiempo supo Mr. R. A. Matthews que yo ya había dejado a Mr. Brown y me escribió que me fuera para Washington, que creía que mejor podría aprender [otro] oficio allí.

CAPÍTULO 11

Cuando volví a Washington me pensé que sería mejor aprender el oficio de sastre que el [de] zapatero. Pues en aquel había tenido mal éxito. Hablé con un sastre que se apellidaba Mudd, y me fui a trabajar con él de aprendiz por tres años, dándome la comida y cama.

Este hombre tenía familia que se componía de su esposa, un niño como de ocho años y una niña de seis y una señorita parienta de su esposa. Eran católicos todos, pero yo nunca los ví que fueran a la iglesia. Lo que sí veía era que iban a los bailes, y [a] los teatros y circos. Este hombre tenía dos o tres oficiales siempre trabajando. [A] su esposa le gustaba mucho los bailes y la sociedad mundana. Tenía dentadura falsa y yo no sé cómo se le des[com]ponían tanto, porque muy a menudo me mandaba con el dentista a componer los dientes. Siempre me traían en mandados y como yo ya era evangélico, no me gustaban sus costumbres. Mr. Mudd echaba muchas mentiras a sus clientes y a veces los traía en vueltas con el trabajo que les hacía.

Me acuerdo de un joven que tenía una panadería, y entregaba pan en un carro. Ya Mr. Mudd le debía una cuenta algo grande. Un día [el panadero] le trajo una capa muy fina y le dijo que quería que le hiciéramos [una] levita de ella. Él se comprometió a hacerla, convinieron en [un] precio, y le tomó las medidas. Y le dijo el joven que la quería para la Noche Buena. Me parece que era a fines de septiembre. El panadero siguió dándole leche a cuenta del trabajo que le iba a hacer. Se pasó como tres semanas y Mr. Mudd no daba principio al trabajo de la levita. Un día le preguntó el joven si ya había comenzado la levita y él le contestó que sí, que no tuviera cuidado, que estaría lista para cuando la necesitara.

A mí me zumbaron las orejas al oír esa mentira tan descarada y, luego que se fue el joven, [Mr. Mudd] se puso a reír, en lugar de darle vergüenza. Después de eso el joven preguntaba todos los días, y la contestación era que ya le faltaba poco para estar lista la levita. Como una semana [antes] de la Noche Buena, se escondía Mr. Mudd cada vez que el joven venía [y] preguntaba por él. Y los oficiales le decían que había salido, mintiendo también ellos, pues él estaba en el cuarto detrás de la oficina. A mí me chocaba tanto aquello, que ya me pesaba haber comenzado a trabajar allí.

Los domingos iba yo a visitar a mis amigos del Hotel United States. En uno de esos días me dijeron en el hotel que un señor mexicano me buscaba, y [que] me había dejado su tarjeta con su dirección. El mismo día fui a buscarlo y lo hallé. Era un joven como de veinte años, nativo de Santa Fe, Nuevo México, que conocía a mi familia y había oído decir que allí estaba yo. Era como cuatro años mayor que yo, y había ido a Washington con don Juan Manuel Gallegos que había sido sacerdote de la Iglesia Romana. Yo me acuerdo [de] haberlo conocido. Había dejado el sacerdocio no sé por qué. Y el pueblo lo había elegido miembro del congreso de los Estados Unidos, y estaba allí como representante del territorio. Y el joven José Sena había ido con él con el fin de ir al colegio en Washington. El padre de este joven se llamaba Juan Sena. Me acuerdo que tenía un comercio en Santa Fe y eran muy bien conocidos de nuestra familia.

Pero ya me voy olvidando de Mr. Mudd y la levita del panadero. Como he dicho, el joven preguntaba todos los días por Mr. Mudd y la levita, pero él estaba escondido, y nadie sabía nada de la levita. Mientras todos sabían y mentían, el día 24 de diciembre en la tarde vino el joven y dijo: "Díganle a Mr. Mudd que por la mañana quiero mi leva sin falta". En la mañana vino y preguntó por Mr. Mudd y recibió la misma contestación de siempre, y [el joven] dijo: "Díganle a ese señor, que digo yo que es un pícaro sinvergüenza, que se coja todo lo que me debe por pan y mi capa, pero que no volveré a comer más de mi pan". Nomás se fue el joven, y salió el [sastre] riéndose. A mí me dió tanta vergüenza y coraje, que ese mismo día me fui y le dije que no quería enseñarme ese oficio. Y me fui al United States Hotel.

CAPÍTULO 12

El señor Gallegos y José Sena me habían propuesto que me fuera para Santa Fe con ellos cuando se acabara el término del congreso, pero yo no tenía recursos para irme con ellos, ni quería pedirles prestado porque sabía que llegando a Santa Fe, tendría que desquitárselos con trabajo y no quería volver bajo esas circunstancias a mi tierra. Pero en esos días leí en un periódico que se llamaba *The Baltimore Sun*, un anuncio donde decía, "Se necesitan cincuenta jóvenes para que vayan a Nuevo México, Texas o California". Me informé que estos jóvenes los necesitaban para soldados. No sé cómo se me vino la idea de escribir al Secretario de Guerra que era entonces Mr. Jefferson Davis y, como no recibí contestación a los tres días fui yo mismo a la oficina y pedí permiso para hablar con el Secretario de Guerra.

Me recibió cortesmente y me señaló un asiento. Cuando le traté del asunto, me dijo: "Those appointments are not open yet". Y yo le dije que no sabía que se había qué esperar algún nombramiento para ser soldado. Entonces me dio la siguiente recomendación: "If the bearer wishes to enlist, I recommend him to the nearest recruiting office which is Baltimore".

El día siguiente tomé el tren para Baltimore y muy pronto hallé la oficina, pues traía el periódico conmigo. En la oficina había un capitán y un sargento. Le presenté la recomendación al capitán y me preguntó mi edad. Le dije que [tenía] diecisiete años y me dijo: "No tienes la edad y además de eso, no das tú el alto que la ley requiere". Entonces le dije:

"Muy bien, adiós" y me salí de la oficina. Y oí que le dijo al sargento: "Sergeant, call that boy back. I want him for my company."

Entonces me preguntó si tenía tutor y le dije que sí, pero estaba en Washington. Me dijo que si podía ir a sacar su consentimiento para entrar de soldado. Le dije que no podía ir, porque no tenía dinero para mi pasaje. Entonces él me dió el pasaje y una notita para Mr. Matthews.

Esa noche me le presenté a Mr. Matthews en Washington. Él no tenía idea de lo que yo andaba haciendo. Le presenté la nota del capitán. "¿Por qué quieres entrar de soldado?" Le dije que quería volver a mi país y ésta era una oportunidad para hacerlo. Me dijo: "Santiago, yo no quiero quitarte tu intención. Quiero que te hagas hombre de por sí, como te lo he dicho otras veces". Me estuvo dando muchos consejos. Le escribió una nota al capitán Palmer, que así se llamaba y otro día volví a Baltimore.

Asenté plaza de soldado por cinco años, el día primero de abril de 1855. El capitán me dijo que me quería para clarín de su compañía, que estaban reclutando para organizar un regimiento. Y me mandó a Ft. McHenry, cosa de tres millas de Baltimore en el Chesapeake Bay. Este fortín que [estaba] rodeado [por] esa laguna, que sólo tiene entrada por tierra por el lado de la ciudad. Allí había dos compañías de artillería. Todo el trabajo que yo tenía era salir todos los días a aprender las llamadas de clarín que se usaban en el ejército. En un mes y quince días, yo ya sabía todas las llamadas y [los] toques de ejercicio.

Estuve allí hasta el mes de agosto. Mi sueldo se me pagaba cada mes. En el mes de agosto, no me acuerdo qué fecha, me llamaron para Baltimore como el día 10, y otro día salí con una partida de reclutas para Newport Barracks, Kentucky. Esta queda frente a Cincinnati, Ohio. Aquí estuve como un mes sin más qué hacer que ir a la escolta todos los días con los demás clarines y tambores. Era una guarnición de infantería, pero allí nos llevaron aunque éramos destinados para caballería.

Durante el tiempo que estuve allí trajeron como diez reclutas que asentaron plaza en Cincinnati, que se comían de canallas, o sea malvados, de ese pueblo. Como a mediado de septiembre nos embarcaron en un barco de vapor que navegaba en el Río Ohio. Ese día se lució la gavilla de reclutas de Cincinnati. Venían todos borrachos y queriéndose pelear con todos hasta que el teniente Low, que venía al mando de la tropa, los puso a todos presos con guardias dobles.

El día que pasamos por Louisville, Kentucky, en lo que estaba el barco parado, recibían pasajeros, y [la tripulación] descargando y cargando carga, los canallas como ya los habían quitado de la prisión, armaron pleito con algunos del barco y en un momento se formó [un] tumulto terrible que parecía que íbamos a perecer todos. Pues aquellos hombres acometían contra todos los del barco y los pasajeros.

Yo no hallaba dónde meterme. Me subí al segundo piso, y al subir la escalera, ví a un compañero mío que se llamaba Metzger, defendiéndose debajo de la escalera de uno de los malvados que se llamaba Swift. En esto el capitán del barco dió orden a su gente [que] sofocara el tumulto, y nomás se dió esta orden, y los negros, que eran los que trabajaban, se les echaron encima a los malvados y, en muy poco tiempo, los vencieron con palos y fierros, [y] con lo que traían en la mano.

Algunos de ellos trataron de escaparse del barco por la andamiada que habían puesto para descargar y cargar flete. Uno de los malvados iba a media andamiada cuando lo alcanzó un negro y le dió un leñazo en un lado de la cabeza y lo hizo caer como diez pies en la agua. Y en seguida, brincó sobre él diciendo: "No te dejaré ahogar, talísimo". Y lo levantó en el hombro y lo sacó a tierra. Los demás revoltosos muy pronto se rindieron y fueron puestos presos debajo de la guardia.

CAPÍTULO 13

Sería como a fines de septiembre o principios de octubre que llegamos a Jefferson Barracks, el lugar donde se había de organizar el segundo regimiento de caballería de los Estados Unidos. El regimiento se componía, en ese tiempo, de diez compañías de ochenta y cinco hombres cada una, con un capitán y dos tenientes. Cada capitán había trabajado para reclutar hombres para su propia compañía. Como a los ocho días de estar en Jefferson Barracks, el comandante del regimiento sacó a todos los clarines al campo. Éramos veinte, dos para cada compañía. Nos estuvo examinando por cosa de tres horas. Nos formó y nos hacía tocar las llamadas o marchas y [los] toques que se usan en el ejército. Y como yo había estudiado más que todos, sabía más y me nombro clarín mayor y me dió cargo de todos los clarines. La plana mayor del regimiento se [componía] del coronel A. S. Johnston, R. E. Lee, teniente coronel y el mayor Hardee y el teniente Hood, ayudante. Los capitanes de las diferentes compañías [eran]: Compañía A, Capitán Van Dorn, Brevet Major; Compañía B, Capitán Smith; Compañía C, Capitán James Oakes; Compañía D, Capitán I. N. Palmer; Compañía E, Capitán Stoneman; Compañía F, Capitán Bradford; Compañía G, Capitán Travis; Compañía H, Capitán White; Compañía K, Capitán Ives. Y no me acuerdo de los nombres de los demás y tal vez me he equivocado en poner los nombres mencionados en las compañías correspondientes, pero son los nombres de los capitanes que me acuerdo.

El día 27 de octubre salió el regimiento de Jefferson Barracks para Texas. En la marcha atravesamos el territorio de los indios. Allí conocí a las siguientes tribus de indios: Los Choctaws, Chickasaws [y]

Cherokees. Estos indios estaban muy avanzados en civilización. Tenían buenas labores, buenos ranchos, buenas escuelas y algunos colegios. Algunos de ellos tenían negros esclavos, que no hablaban el inglés. Para mí era cosa muy curiosa oír a los negros hablar en la lengua de los indios y no entender el inglés. Como había visto miles en el Sur que no hablaban otra cosa [que inglés].

Algunos de los indios que ví eran muy ostentosos. Llegué un día a una casa donde estaba un indio viejo sentado [en] el corredor en una silla mecedora. Fui con el fin de comprar leche o huevos. Me dijo que pasara y me sentara en lo que me traían lo que buscaba. Yo, deseoso de entablar conversación con él, le pregunté cómo se llamaba y me contestó: "George Washington" y fueron todas las palabras que [le] pude sacar.

Tardamos como dos meses para llegar a Texas. Pasamos el Río Colorado que es la línea entre Texas y la nación de los indios. En un pueblito que se llama Preston, que está del lado de Texas, allí había un chalán y en esa pasó todo el regimiento. Tardaríamos como ocho días para llegar a Fort Belknap. No me acuerdo de la fecha, pero ya era después de Noche Buena. Allí había cuartos, pero no suficientes para todo el regimiento. A mi compañía le tocó cuartel y a dos o tres más. Los demás se acamparon en tiendas de campaña de las cuales veníamos bien abastecidos. Unos cuantos días antes de llegar a [Fort] Belknap, una noche nos pescó un norte muy fuerte y muy frío. Muchos de los que no habíamos experimentado un norte de ésos lo extrañamos mucho. Se oía que decían los soldados: "Texas Norther, boy."[6]

Sería como a las nueve de la noche, y como había muchas lumbres en el campo,[7] volaban las brasas por todo el campo y tuvimos que apagar las lumbres. En lo que apagábamos las lumbres se me voló mi sombrero, y no lo pude hallar. Yo sufría de un dolor de muela y esa noche estaba en un tormento. Como a media noche, fui a la tienda del doctor con una desesperina [sic] que parecía que estaba rabioso. Me paré en la puerta, y empecé a gritarle al doctor. Me preguntó "¿Qué quieres?" y le dije, "Estoy loco con el dolor de muela". Me dijo que fuera con el boticario.

Fui a la tienda de éste y me dijo: "Anda con el doctor". Le dije, "Ya fui con él y el me mandó contigo".

"¿Pues qué quieres que te haga a estas horas y con tanto frío?"

Le dije: "¡Hazme algo, porque me muero!"

Se levantó y me echó en una botellita una cantidad [de] creosoto. Y se me quitó el dolor, pero me quemó todo el lado de la muela, por dentro y fuera. Por mucho tiempo tuve el moretón y sufrí de la quemada. Por algún tiempo no pude tocar el clarín.

Sería en el mes de abril cuando nuestra compañía y otras dos más marchamos de Belknap para Fort Mason. Allí había cuarteles, pero estaban ocupados por dos compañías de nuestro regimiento que se habían venido desde el invierno. En ese tiempo se le ofreció al capitán de mi compañía hacer un viaje para San Antonio con su familia.

CAPÍTULO 14

Había un cabo en la compañía que se llamaba Prusia.[8] El capitán lo nombró a él y [a] una escolta de cinco hombres para que lo acompañaran en su viaje. Salimos de Fort Mason después de medio día. Yo era uno de la escolta. Los soldados salieron con sus caramayolas llenas de whiskey. Esa noche llegamos al Río del Llano, al rancho de un alemán que se llamaba Martin. Éste era matancero que entregaba carne a las tropas en Mason. El capitán y la familia pasaron en la casa y nosotros pusimos el campo como [a] cien varas de la casa.

Antes de acostarnos se suscitó una discusión entre los compañeros [de] si sería necesario poner centinela o no. Unos arguían que no había peligro de indios, y entre éstos era el cabo. Yo estaba a favor de que pusiéramos centinela, pues siendo soldados era nuestro deber. Y yo estaba más sobrio que todos. Llegó a tanto la disputa que ya mero se peleaban, y el cabo era el más acalorado.

Como era de noche y no teníamos más que la luz de la lumbre, no vimos al capitán que se había venido a ver qué pasaba. Yo fui el primero que lo ví, pero según él dijo, ya hacía rato que estaba oyendo todo. Luego que lo ví, dí el orden: "Attention!"

Todos nos paramos firmes según el costumbre militar al acercarse un oficial. El capitán dirigiéndose al cabo, dijo: "He estado [escuchando] vuestro proceder por algún rato, y siento mucho que después de haber escogido determinadamente mi escolta para este viaje, he hecho un equivoco. Yo creía que el cabo Prusa era uno de los mejores soldados de mi compañía y siento mucho ver que ni quiere poner centinela para cuidarse del peligro de los indios, pero ni para cuidar de los bienes del gobierno que trae a su cargo". [Se refería a] los seis caballos que montábamos y [las] seis mulas del carro en que venía el equipaje y forraje y provisiones y las dos mulas del coche en que venía el capitán y su familia. Entonces dirigiéndose [a mí], el capitán dijo: "Tafolla, tómese usted carga de esta escolta", y se devolvió para la casa. Yo como ya les había dicho que la haría de primer centinela, les dije que se acostaran.

Serían las diez de la noche [cuando comencé y] como a las doce desperté a otro. Como a las dos, otro. A las cuatro, uno que se llamaba Muth. Éste era uno de los que se oponían a velar. Le dije: "No te sientes, porque te quedas dormido". Había muchos marranos en el rancho y toda la noche tuvimos que estarlos corriendo. Entre los trastos del capitán, venía una silla mecedora y la habíamos apeado. Las provisiones estaban todas alrededor del campo, como también un saco lleno de maíz para los caballos y las mulas. Muth arrimó todo tan cerca como pudo alrededor de la silla, y se sentó en ella y se durmió.

Como a las cinco desperté yo, y había como cincuenta marranos en el campo. [Se] habían comido todo el maíz y las provisiones, y traían a Muth con todo y silla vuelta y vuelta. A ese tiempo un marrano grande se metió debajo de la silla y la volteó con todo y Muth. Yo andaba con un palo en la mano corriendo [a] los marraños. Al golpe despertó Muth asustado, diciendo: "What's the matter?" Yo le contesté: "Ya quedamos sin provisiones y sin forraje para las bestias".

Cuando el capitán vino, halló las nuevas. Yo le platiqué todo. Dijo: "Bueno, vamos a llegar a Fredericksburg para comprar provisiones". Nomás el café nos habían dejado los marranos. De allí a Fredericksburg había como veinticinco millas y vinimos a dormir a Cíbolo. El cochero que traía al capitán, el señor Jefferson, también era soldado.

Otro día llegamos a San Antonio. Paramos[9] en el corral del gobierno, que estaba contra la capilla que se conserva todavía por el Alamo building. El lugar que se llama ahora la plaza del Alamo Plaza [sic] era entonces un chaparral. En donde está ahora el edificio de la ópera y todos esos edificios al lado norte y sur, era puros jacales techados de zacate. Esto fue en julio de 1856. El lugar que ocupa ahora el Hotel Menger era todo chaparral de mezquite. Más tarde en ese mismo año comenzaron a cavar los cimientos para el gran Hotel Menger. Esa fue la primera vez que yo conocí San Antonio.

Ese mismo año, el capitán Palmer recibió orden de establecer Camp Verde en un arroyo de ese nombre en Kerr County [a] once millas de

Kerrville, "the county seat", y [a] doce millas de Bandera Town. Ambos pueblos se componían de unas cuantas casas y un molino de rajar madera. En este campo se edificaron cuarteles para los oficiales y para los soldados.

Durante ese tiempo el capitán me eligió a mí para que corriera el correo de allí al camino de Fredericksburg y San Antonio, [y] como veinte millas [de allí] a un rancho de un alemán que se llamaba Zink. Este viaje lo hacía yo todos los viernes. Esos eran tiempos peligrosos porque los indios entraban y salían robando caballos y matando a todos los que encontraban. No sé porqué me dió ese trabajo el capitán, tal vez sería porque yo era liviano para el caballo. Muchas veces en tiempo de calor hacía el viaje de noche. Siempre sabía yo de las novedades de los indios. A veces pasaban después y a veces antes que volviera al campamento, pero nunca los encontré.

Pero el capitán por algo me escogió. Una noche después de [pasar] la lista de [las] seis, se puso el capitán enfrente a la compañía y dijo: "Quiero un hombre que vaya a llevar una carta a Castroville esta misma noche. ¿Quién quiere ir? ¿Quién sabe el camino?" Nadie contestó. Yo le dije: "Yo no sé el camino, pero me atrevo a ir". Y él dijo, "I know you will. Get ready and come to my office for the letter".

Serían como las nueve de la noche cuando salí de Camp Verde. Llegué a la Bandera como a las once. Allí desperté a unas gentes americanas, y les dije que era soldado de Camp Verde, y que iba de correo para Castroville. Salió un hombre con una carabina en la mano, porque en esos tiempos no podía uno salir de la casa sin armas por temor a los indios y después que se satisfació que yo iba de buena fé, me dió la dirección del camino.

CAPÍTULO 15

Allí atravesé el Río de Medina. Como a diez millas de allí volví a pasar el río y [a] como tres millas de allí llegué a un lugar que le decían Mormon Town. Había una colonia de mormones allí, de cosa de unas diez o doce familias. Sería como a la 1:30 [a.m.] cuando llegué. Ví luz en una casa y me dirigí a ella.

Al acercarme, oí música y ví que estaban bailando. Me pasé como a diez pasos de la casa y dí el "Hallow"[10], y al momento se paró la música y salieron como seis hombres con las armas en la mano. Luego que me reconocieron y supieron mi negocio, me invitaron que pasara y comenzaron otra vez a bailar.

Había como unas quince señoras y señoritas y otros tantos hombres, y todos parecían que estaban gozosos y felices. Me invitaron a un asiento y un señor anciano que parecía ser el jefe de todos se sentó junto a mí y me dijo: "Estamos festejando un casamiento que se verificó entre nuestras familias", y me dió la bienvenida. Me preguntó si quería bailar. Yo le dí las gracias y le dije que tenía que estar en Castroville para el salir del sol. Mandó que le dieran maíz a mi caballo, y me llevó a otro cuarto donde tenían una mesa con todos los manjares delicados que se podía desear. No ví licor de ninguna clase ni señas de que nadie anduviera tomado. Desde esa noche me formé la opinión de que los mormones eran buena gente. Estuve allí como una hora. Me invitó el señor que la hacía de jefe a que llegara en mi vuelta.

Llegué a Castroville como a las seis de la mañana. La carta que llevaba era para Mr. Vance. Pregunté por él y pronto hallé la casa, pero me dijeron que andaba en San Antonio y que no vendría hasta después de mediodía. Yo no llevaba dinero para mi comida o la de mi caballo. Lo amarré en una sombra de un encino muy grande que había allí y lo dejé ensillado, mientras yo me fui a dar vuelta al pueblito. Para esa hora ya tenía mucha hambre.

Como a la una vino el señor Vance y me dió una carta en contestación a la que le traje. Entonces le dije que quería que me diera de comer, para mí y para mi caballo, que yo no tenía dinero, pero que le cobrara al gobierno, pues yo andaba en negocios del [gobierno.] Entonces me dijo que la carta que le había traído era con respecto a un contrato de maíz que él tenía con el gobierno y que él creía que ya lo estarían entregando en Camp Verde. Mandó que me dieran de comer [a mí] y a mi caballo.

Y serían como las cuatro de la tarde cuando salí de vuelta. Como a las nueve de la noche pasé por Mormon Town, y esta[ba] todo muy quieto. Entre Mormon Town y Bandera dormí cosa de tres horas. Cuando llegué a Camp Verde, ya las carretas habían traído maíz. El capitán podía haberme ahorrado ese viaje, pero como tenía soldados podía mandarlos cuando él quería. Ahora que conozco el evangelio me acuerdo de lo que le dijo el centurión al Señor: Mateo 8:9.[11]

[En] otra ocasión me despachó el capitán para Fredericksburg de travesía con el apuro de que fuera a traer un doctor. Sin conocer yo el terreno, otra vez me mandó a traer a un juez de paz al Río Guadalupe, cerca de Kerrville, para que viniera a tener un Inquest: [a] levantar el acta de uno que habían matado cosa de una milla de Camp Verde. Así me usaba en casos semejantes. Yo creo lo hacía porque yo era mexicano.

En un[o] de mis viajes que hice de correo, me pescó un norte muy frío, lloviendo y candelillando.[12] Llegué al campo cosa de las seis p.m. Yo vivía en el cuartel de los sargentos y cabos. Había seis hombres en el cuarto, y no tenían lumbre. Yo luego fui y traje leña y puse lumbre y empezaron algunos de ellos a decir que si no hubiera venido yo, se hubieran helado de frío.

Yo me había lavado la cara y estaba frente al espejo peinándome y diciéndoles que todos ellos eran flojos, por eso no ponían lumbre. Y uno de ellos que era cabo y se llamaba Dennis me dijo que si no me callaba la boca me echaba fuera del cuarto a patadas. Le contesté que no era hombre para hacerlo y que si se atrevía, le daba un balazo.

Yo andaba muy enojado, porque se estaban burlando de mí, después de haberles puesto lumbre. Entonces Dennis se vino hacia mí como a darme un golpe por detrás, y yo lo ví en el espejo. Al momento saqué la pistola que la traía fajada, con intento de darle un tiro. Pero al momento me agarraron entre todos y [me] llevaron y me pusieron preso en la casa de la guardia y le pasaron el informe al capitán de lo que pasaba. Esa noche dormí preso.

Ese mismo día había venido el pagador. Yo tenía la esperanza de que el capitán me echara fuera otro día a la hora de montar guardia, como era costumbre hacer con los presos que no eran criminales. Pero no lo hizo, sino que me mandó llevar a su cuartel, y allí me habló.

Me dijo que si le pedía perdón al cabo Dennis, me daba libre, y yo le dije que yo me consideraba con derecho a defenderme y lo había hecho en defensa propia. Conocí que él lo que quería era que me humillara, lo que justamente debía de haber hecho.

CAPÍTULO 16

Pero yo era muy orgulloso, había perdido mi religión, y Satanás se había aprovechado de esta ocasión. El capitán me dijo: "Si no le pides perdón al cabo Dennis, tendrás que estar preso por algún tiempo, y al fin serás juzgado por una Corte Marcial". Le dije: "Si usted me tiene preso hasta que cumpla mi tiempo, que son tres años y medio, no le pediré perdón a Dennis".

Había una noria de cosa de treinta pies de hondo en medio del corral, que se componía [de] como doscientas varas en cuadra. En un lado estaban las caballerizas de los caballos; en otro, la de los camellos, en otra la de las mulas. En la otra cuadra estaba la casa de la guardia y la comisaría. En medio del corral comenzaron a hacer una noria, y no la habían acabado. Tenía como treinta pies de hondo y allí ponían presos a los soldados que se emborrachaban, con el fin de que no consiguieran licor en que estaban presos.

Pero ellos [los soldados borrachos] ya habían inventado un modo de conseguirlo. Se habían puesto de acuerdo con algunos de los que habían salido de allí para que en las noches les trajeran vino. El que traía el vino daba un silbido y tiraba una piedrita para dentro de la noria. Y al momento se pasaban todos y agarraban una frazada de las puntas y la alzaban sobre sus cabezas. Y el de afuera tiraba una botella o una caramayola de whiskey. Y como era de noche, las guardias no veían nada.

En esa noria me echaron a mí. Nomás una noche me tuvieron, y otro día me mandó sacar el capitán y me dijo: "Tafolla, no quiero verte en la noria con los borrachos. Ni creo que tú quieres estar allí. Quiero que me prometas que no buscarás pleito con el cabo Dennis, y te daré libre". Yo le había dicho que el día que saliera libre, le demandaría satisfacción a Dennis. Le dije que no le podía hacer esa promesa. Entonces me despachó otra vez a la guardia, pero ya no me echaron a la noria.

Como en esos días nos habían pagado, todos los soldados se ocupaban de jugar, procurando ganarse el dinero unos a otros. Yo era el único preso que había en la casa de guardia. Los demás estaban en la noria, como he dicho. Y estaba oscureciendo, y el cabo de guardia no hallaba a los centinelas que debían de tomar el primer puesto que comenzaba a las seis de la tarde. Todos estaban en los cuarteles en la tarreada.[13]

Me dijo el cabo: "Jim, ai te encargo la caballada en lo que yo voy a traer al centinela". Al momento me entró la idea de montar un caballo y desertarme. Fui y agarré un caballo y lo llevé al otro lado del arroyo, que estaba como [a] trescientas varas. Lo amarré allí, y me devolví a traer mis frazadas que las tenía allí en el cuartel de la guardia, porque era tiempo de frío y tenía que caminar toda la noche. Cuando volví a donde había dejado el caballo, ya se había soltado y se había ido. Dejé allí mis frazadas y volví al corral a traer otro caballo. Estaba desatándolo cuando oí que venía el cabo Brown, que así se llamaba, arrastrando la espada. Entonces empecé a hablarles a los caballos, como que los andaba aquietando. Al mismo tiempo llegó el caballo que había dejado, del otro lado del arroyo. El cabo me habló y me preguntó qué pasaba. Le dije que un caballo se había soltado y lo iba a amarrar. Me pidió favor que cuidara otro rato y se fue a buscar [a] su gente. Ya para esa hora estaba bien oscuro. Yo me llevé el caballo, le eché las frazadas encima y monté en él. Caminé toda la noche sin saber el camino, aunque tenía una idea del [rumbo.]

Como a las cuatro de la mañana llegué al arroyo de San Jerónimo, al rancho de don Manuel Hernández. Allí estaba Federico Rivas, que me había conocido ya en Belknap y me había visto en Campo Verde preso el día anterior. Luego que me vió comprendió que venía desertado, y le dije que no era mi idea desertarme, sino presentarme al coronel del regimiento y pedirle que me hiciera justicia. Este amigo me prestó una silla y ese día llegué a San Antonio como a las once de la mañana.

Fui a la oficina del coronel y le platiqué todas las circunstancias, y él me dijo: "Tafolla, I am sorry you got into trouble. Lleva tu caballo a la caballeriza del gobierno, y paséate en la ciudad. Yo creo que te seguirán. Hoy mismo escribiré al capitán Palmer [diciéndole] que te has presentado aquí".

A los dos días vino uno de los guías que se llamaba Jim Claude, y el sargento Hemphill y dos soldados. Se presentaron en la oficina del coronel y supieron que yo estaba en la ciudad. Había un cierto hotel donde parábamos los soldados, y allí fueron a pasar. Estaba yo cenando cuando entraron. Me saludaron muy contentos, como compañeros que éramos. A los dos días salimos para Campo Verde.

Cuando llegamos, fue el sargento a dar su reporte y le dijo el capitán que me entregara a la guardia. Esa noche me echaron a la noria otra vez. Otro día me mandó llevar el capitán a su oficina, y me dijo: "Tafolla, yo hice todo lo que pude por librarte del castigo y, como tú sabes, te he tratado siempre con mucha consideración, más que a ninguno de mi compañía, y en lugar de tomar mis consejos, tú te has hecho criminal. El castigo de un desertor es muy severo, como tú lo sabes. Pero ahora quiero darte otra oportunidad. Si tú me prometes que no buscarás pleito con el cabo Dennis en los cuarteles o estos terrenos de Campo Verde, yo hablaré por tí a las autoridades militares, y si en alguna ocasión tú y el cabo Dennis se encuentran fuera del campo y quieren arreglar su asunto, pueden hacerlo. Pero no en los cuarteles o en los límites de este campo militar, porque como tú sabes, mi compañía siempre ha tenido buen nombre, y yo no creo que tú quieras ser causa de que lo pierda". En verdad era la compañía que tenía el mejor nombre en el regimiento. Yo le pedí perdón al capitán, por haber sido tan estúpido y le prometí hacer como él me decía. Ese mismo día me dió libre.

CAPÍTULO 17

Como a los ocho días después de esto se ofreció que saliera un pico de soldados a seguir los indios que habían entrado y habían robado la caballada del Dr. Nowling que vivía en el Río de Guadalupe, como [a] treinta millas de Campo Verde, para abajo. A mí me gustaba salir en esas campañas y pedí permiso al capitán que me dejara ir. Y como mi pistola me la habían quitado desde que tuve el choque con Dennis, me dieron otra en muy mal orden y un rifle de los que se usaban en el ejército en ese tiempo.

Llevamos por guía a J. P. Rodríguez que era muy perito para seguir la huella de los indios. En las cabeceras del Río de Guadalupe les agarramos la huella de los indios. El guía tenía una mula del gobierno que montaba siempre que seguía indios y tan diestro estaba él como la mula [en seguir] la huella. Él tenía mucha experiencia en seguir indios. Ponía la mula en la huella, y nomás iba mirando adelante como diez o veinte varas. Nosotros no veíamos nada de huella, pero él veía adelante el zacate o las yerbas dobladas o las piedritas volteadas y así nos llevaba al trote.

Como a los seis días pasamos como a seis millas de Fort McCavett. Ya algunos de los caballos habían perdido las herraduras. Este fort queda en las cabeceras del Río de San Saba. El sargento de orden de nuestra compañía era el jefe de la gente, y determinó que llegáramos a Fort [McCavett] para herrar los caballos, que ya varios iban mancos, porque el terreno que habíamos andado era muy pedregoso. Llegamos al fortín como a la una p.m., y salimos como a las cinco de allí. Sacamos también provisión. Llegamos al lugar donde habíamos dejado la huella ya oscureciendo, y allí dormimos.

Yo nunca he sido supersticioso, pero esa noche soñé un sueño que habíamos alcanzado los indios. Habíamos peleado con ellos. Que un indio se había cortado de los demás, y que yo y John Martin nos habíamos ido tras de él. Que lo alcanzamos en una mota y lo herimos y lo tomamos prisionero, y lo llevábamos por delante. Cuando de repente, con las dos manos amarradas por delante, le arrebató a Martin una daga que él [Martin] cargaba en la faja de la cartuchera y se lo embasó cerca del corazón. Al mismo tiempo me acometió a mí y sin darme tiempo de defenderme, me embasó la daga [en] la ingle, abajito de la faja de la cartuchera. Sentí el dolor, grité y desperté. Martin, que estaba durmiendo conmigo, también despertó al grito y me preguntó qué pasaba. Yo le platiqué el sueño, y se rió. Ya estaba aclarando y nos levantamos todos, hicimos nuestro café y seguimos la huella.

Ese día como a las cuatro de la tarde, uno de los espías que iba a la derecha nos notificó [que] allí cerquita estaban los indios acampados. Al mismo tiempo los vieron los compañeros que iban más [a]delante. No eran los que íbamos siguiendo. Era otra partida que venía entrando a robar caballada y [a] matar a la gente que les estorbara en su paso, como lo hacían siempre. Sería[n] como quince indios, y estaban comiendo. Traían nomás tres bestias —una mula y dos caballos. En estas bestias traían su equipaje— varios cueros de cíbolo curtidos, muchas teguas (zapatos), gamusas. Y por todos los preparativos que traían, comprendimos que venían de expedición.

Habían soltado las bestias en un llano que había a la derecha, y un indio había salido a traerlas para el campo. Cuando el indio que venía con las bestias nos vió, dió el alarme y dejó las bestias y se juntó con sus compañeros. A la izquierda del campo había un monte muy fuerte y por medio de él pasa un arroyo de agua que nacía del ojo de agua del Quikapoo Spring[14], como le dicen en inglés.

Los indios se metieron en este monte y de allí nos empezaron a hacer fuego con flechas y con armas de fuego, pues venían bien armados. Fue tan grande la sorpresa que les dimos que algunos dejaron las armas junto [a] la lumbre donde estaban comiendo. Uno de ellos salió a llevar una carabina que había dejado y antes de entrar al monte cayó muerto pues los soldados habían desmontado y comenzaron a hacer fuego.

La idea de los indios era quitarnos la caballada, y se dividieron en tres partidos. Unos nos atacaban por la derecha, y otros por la izquierda, y otros al frente. Y como estaban en el monte, no los veíamos mas que cuando nos arremetían. Al darles nosotros el primer asalto, el guía, que era el único que había peleado con los indios antes, dijo al sargento: "No deje que los hombres se metan al monte a caballo, porque no sale ni uno vivo". Entonces el sargento dió orden que nos tiráramos a pie y encadenáramos los caballos como habíamos aprendido ya en el ejército, de manera que un hombre podía tener veinte caballos.

Yo me apeé de mi caballo y lo dejé suelto. Nomás le tumbé la rienda, porque era muy mansito. Los indios hicieron varios impulsos a meterse entre nosotros y la caballada para quitárnosla. En una de ésas, se ahilaron unos indios por el lado donde yo estaba, determinados a cortarnos de la caballada. Entre tanto, los otros hacían lo mismo a la derecha, porque yo estaba a la izquierda y estábamos formados en tiradores.

A ese tiempo oí la voz del guía que le dijo al sargento: "Monten los caballos o nos los quitan". Yo creí que todos lo iban a hacer. Yo volteé y ví que mi caballo me había seguído y estaba cerquita de mí. Tiré el rifle con que estaba peleando, monté mi caballo, saqué la pistola y me fui a encontrar a los indios que venían ahilados. El que venía delante se compuso el chimal y se quedó esperándome, brincando y gritando su gritito de guerra, como ellos acostumbraban. Los demás indios se devolvieron.

CAPÍTULO 18

Yo le tiré el primer tiro como [a] veinte varas [de] lejos, pero como iba al galope, me acerqué pronto. El segundo se lo tiré como [a] diez pies y el tercero creí ponerle la bala en la cabeza. Pero al montar la pistola, el cilindro se atoró con el casquillo del cápsul, y cuando quise ya estaba sobre el indio. Y como él estaba resguardado con el chimal, no le pegó ni una bala. Pero lo que hice fue darle con la pistola en la cabeza. Al mismo tiempo sentí el flechazo en el abdomen y sentí el dolor lo mismo que lo había sentido en el sueño la noche antes. El indio se enderezó, esperando verme caer del caballo. Y como vió que no caía, hizo una demostración de admiración y se echó el chimal a la espalda, y se metió al monte.

Yo detuve mi caballo y volví sobre el indio. Y ví que los compañeros ninguno había montado. Uno de mis compañeros, muy amigo mío, que se llama Fred Metzger y ahora vive en Hondo City, me había seguido a pie, y levantó mi pistola (que la solté cuando le dí al indio con ella.) Al mismo tiempo vi que un indio le tiró a John Martin con una carabina y la bala le atravesó el cuerpo, entrándole debajo de la tetilla izquierda y pasándole por los bofes.[15]

Todo esto me hizo pensar del sueño. A mí no me mató el indio porque me había quitado el sobretodo, de los que usan los soldados, y lo había envuelto muy apretado y lo había amarrado en la cabeza de la silla con una correa de vaqueta que tenía para el propósito. La flecha pasó como sesenta dobleces del sobretodo, pues tocaba[16] las mangas, el forro que era muy grueso y lo mas grueso del sobretodo, que es la espalda. También pasó la cartuchera y la faja y mi ropa, y me penetró el cuerpo como media pulgada. Todo esto por la providencia de Dios [quien] me salvó la vida. Yo creo que si el indio me hubiera matado, su intención era montar mi caballo y procurar así quitarnos la caballada.

El guía, J. P. Rodríguez, cuando les dimos el asalto a los indios, lo primero que hizo fué atajar las bestias de los indios y echarlas detrás de nuestros caballos, y allí se estuvo cuidándolos y viendo el pleito. Dice que él veía el polvo del chimal del indio cuando le daban las balas de mi pistola cuando me encontré con él. Esto dice él en su historia que escribió, con respecto a este pleito de indios.

Me acuerdo que el teniente segundo de nuestra compañía así lo mataron los indios durante la "Expedition to the Wichita Mountains". Y fueron cuatro compañías bajo el mando del mayor Van Dorn. El teniente Van Camp se encontró con un indio de la misma manera que hice yo, y el indio le atravesó el cuerpo con una flecha y cayó muerto. Y el caballo corrió solo en rumbo [al campamento] de los indios. Conocí muy bien el caballo que él montaba. Era muy brioso, y tal vez los indios no lo pudieron agarrar.

Después que me hirió el indio me apeé del caballo. Comencé a tirar con mi rifle otra vez. Dos de los compañeros se habían llevado al herido para atrás. A ese tiempo un indio muy alto venía por detrás de unos chaparros con una pistola de las que les llamábamos dragonas, buscando el claro para tirarme en lo que yo estaba cargando mi rifle y preparándome para tirarle a él también. Pero Fred Metzger estaba detrás de los mismos chaparros, y el indio no lo veía y estaba como [a] quince pies de él.

Levantando el indio la pistola para tirarme, le tiró Metzger y le dió en el estómago. Le debe haber atravesado el cuerpo y trozado el hueso del espinazo, porque el indio saltó para arriba y cayó muerto. Yo ya estaba metiéndole puntería al indio, y le dije a Metzger: "Me quitaste mi tiro". Y él me dijo: "I got him". Al momento que cayó ese indio, salieron dos del monte y lo llevaron arrastrando, y tocaron pito y se fueron llorando.

El sargento quería que entráramos al monte para ver cuántos indios habíamos matado, pero J. P. Rodríguez le aconsejó que no lo hiciéramos porque los indios heridos que había allí nos mataban y ya estaba haciéndose muy tarde.

Entre los soldados había uno que se había quedado atrás de los demás, escapándose de entrar al pleito. A éste lo maltrató mucho el sargento por cobarde, y éste era un hombre que en los cuarteles, cuando tomaba whisky, era muy valiente y todos le tenían miedo. El sargento lo acusó de cobardía ante el enemigo, y lo pusieron preso. Y como un mes después, lo juzgaron por una corte marcial y lo echaron fuera del ejército "with a dishonorable discharge, unfit for the service of his country".

CAPÍTULO 19

Luego que los indios se fueron, el sargento hizo a McDonald que cargara todo el equipaje de los indios y lo llevara hasta donde estaban las mulas de carga. Debo de decir aquí [que] el Dr. Nowling y otro americano que se llamaba Moore iban con nosotros y, cuando comenzó el pleito, el doctor se dejó caer de [su] caballo y disparó un fúsil de dos tiros que llevaba. Y en seguida sacó la pistola y les siguió tirando hasta que se le acabaron los tiros. Después se devolvió a donde estaba su caballo y se puso a cargar sus armas parado detrás del caballo. Él era un hombre alto y llevaba unos retazos de frazada colorada que le cubrían las piernas. Los indios tal vez creían que era el capitán. Le tiraron un tiro con carabina y le dieron al caballo en la cabeza detrás de la oreja y cayó muerto. Al momento empezó a quitar su silla y la llevó para atrás, diciendo: “Así hacen estos pícaros cobardes que se meten al chaparral en lugar de salir al llano como los hombres”.

Luego dimos providencia salir de allí para llevar al herido al doctor de Fort McCavett, que era el lugar mas cerca. Ensillamos mi caballo con una silla de los indios. Y nos compusieron estribos a cada uno, y nos pusieron unos cueros de cíbolos. Y nos pusieron a los heridos en el mismo caballo. Salimos del lugar oscureciendo. Le preguntó el sargento al guía si podríamos llegar esa noche a Fort McCavett, y él le dijo que creía [que] llegaríamos al amanecer. Caminamos al rumbo que el guía nos llevaba.

Como a la media noche empezaron los soldados a repelar, diciendo que íbamos perdidos y que en esa dirección nunca llegaríamos al Fort. El sargento se paró y consultó al guía. Él le dijo: “No vamos perdidos. En poco rato vamos a llegar a la laguna donde dimos agua ayer en la mañana". Habramos andado como una milla cuando oímos gritar unos patos. Y dijo él: "Oigan esos patos en la laguna". En efecto llegamos, y dimos agua y seguimos nuestra marcha. Cada rato le preguntaban los soldados al guía: "How far is it to Fort McCavett, Polly?" Y él decía: "Ya no está muy lejos".

Sería como a las 5 a.m. cuando [oímos] el grito, "Who comes there?" Era el centinela de Fort McCavett. Tan pronto como llegamos, se hizo el reporte al comandante y se levantaron las tropas y nos recibieron cordialmente. A mí y al herido nos llevaron al hospital como a las 6 a.m. Vino el doctor y se hizo cargo del herido. Yo venía algo hinchado de la herida, y me dolía, pero quise estarme en el hospital. El doctor me untó un unguento y me puso un trapo. Me decían que los indios emponzoñaban las flechas, y yo eso temía. Pero como la flecha tuvo que pasar por tanta ropa, no le podía quedar nada [del veneno], y creo que así fue.

A las veinticuatro horas de nuestra llegada, murió Martin, y ese día a las 3 p.m. lo sepultamos con honores militares. Todas las tropas y la banda de música nos acompañaron. Otro día salimos. Cuando llegamos a Campo Verde me dijo el capitán: "Tafolla, te escapaste que te matara un indio". Y yo le dije: "Si yo hubiera tenido mi propia pistola, hubiera matado al indio". Porque yo siempre tenía mi pistola en buen orden, pero me la quitaron cuando el pleito con Dennis y me dieron otra.

En el año de 1858 se rumorió la guerra contra los mormones de Utah y nuestro regimiento se puso en marcha para ir a reunirnos con las demás compañías en Camp Cooper, pero a pocos días de haber llegado a ese lugar, vino la noticia de que ya se había arreglado todo, y nuestra compañía y otras dos nos quedamos allí por cerca de dos años.

Fue durante ese tiempo que se organizó la expedición a las Wichita Mountains, y el segundo teniente de nuestra compañía fue nombrado en ella. Como una milla del campo militar estaba lo que le llamaban "the Indian Reservation," que se componía de una tribu de indios comanches. Serían como mil con todo y familias. Estos estaban a cuidado del gobierno. Habían hecho paz y el gobierno los mantenía y les tenía una escuela para sus niños. El jefe de ellos se llamaba "Ketomsey". Estos tenían parientes y amigos entre los broncos que no se querían sujetar al gobierno. Los broncos venían a las habitaciones [de Camp Cooper] y robaban caballadas y mataban gente. Muchas veces los habitantes seguían las huellas hasta la reservación y querían matar a los indios creyendo que ellos eran los malhechores, pero los ladrones se pasaban con el robo. El comandante de la tropa les decía [a los indios] que señalaran a los ladrones y los entregaran, pero ni ladrón ni robo se hallaban allí.

Una vez agarraron los indios a dos indios broncos que vinieron a robar, y los entregaron a las tropas. Estos dijeron que pertenecían a la banda de Buffalo Hump y que éste estaba campado en cierto lugar en las cabeceras del Río de los Brazos. El coronel organizó una expedición como de doscientos hombres y el mismo coronel y el Mayor Thomas salieron a buscar a Buffalo Hump. Llevábamos a los indios prisioneros. Ellos nos iban a enseñar el lugar. Iba también Ketomsey, el jefe de la reservación y veinticinco indios de los mansos. Los indios prisioneros decían que en cinco días llegaríamos, pero caminamos diez días y no hallamos nada. Todos los días nos decían, "Mañana llegamos". Hasta que el coronel conoció que nos iban engañando. Ketomsey todas las noches les arengaba a los indios como una hora.

Había entre ellos un cautivo mexicano que se llamaba Encarnación. Él pertenecía a Ketomsey. Y me decía que las arengas del jefe eran aconsejando a los indios que se hicieran leales al gobierno, y que si hallábamos a Buffalo Hump, y ellos [los indios broncos] querían pelear, que no se metieran. Pero que las tropas no iban con el fin de pelear, sino de persuadirles a que hicieran la paz con el gobierno.

CAPÍTULO 20

En esta campaña iban seis compañías. Todos los días que emprendíamos la marcha, una compañía iba como una milla a la derecha y otra a la misma distancia a la izquierda y dos en el centro. Detrás de estas iban las mulas de carga. Los indios iban como una milla adelante, de exploradores. Toda demostración de peligro se reportaba al centro. Un día los indios avisaron al capitán de mi compañía (que le tocó ir a la izquierda) que habían descubierto huellas de caballos, frescas. Al momento se notificó a[l] coronel, y pararon las tropas en lo que los indios examinaban bien las huellas. A poco rato vino un indio y dijo que eran dos caballos sueltos y los iban a agarrar. A poco los vimos venir corriendo con los dos caballos por delante. Los pasaron por contra la compañía. Eran dos caballos mexicanos muy gordos. Los pasaron por junto [de] nosotros dos veces. Los soldados no podíamos correrlos [para lazarlos], porque había orden muy estricta: que ninguno siguiera a trotas a su caballo[17], y si alguno lo hiciera durante la marcha del día, lo hacían que fuera a pie el resto del día. A mí se me salía el corazón por lazarlos, pero temía desobedecer.

En una vuelta que dieron los indios por contra la compañía tras de los caballos, el teniente de mi compañía le metió espuelas al caballo y los siguió. Yo ya tenía hecho [el] lazo y lo seguí. El sargento gritó: "Halt! Halt!" pero yo no me detuve. Como mi caballo era ligero, en un momento los alcancé. Iban corriendo tan juntos que parecía que iban mancornados. Les eché el lazo a los dos juntos y al momento se pararon. Eran caballos mansos que los indios habían dejado cansados, quién sabe desde cuándo. El teniente me dijo: "Give me one, Tafolla" y se lo di, porque temía el castigo. Y nunca me dijeron nada. A los tres días hallamos dos más, un caballo y una yegua. Los indios los anduvieron corriendo. Y nosotros llegamos a un charco muy grande de agua, y paramos allí.

Desensillamos y soltamos a los caballos, cada compañía por separado como lo hacíamos siempre. Yo le había prestado mi caballo a un compañero, que se le había cansado el caballo. Y era de los de [la] retaguardia que venían atrás con las mulas de carga. Los caballos se les habían ido a los indios, y los dejaron porque sus caballos estaban cansados. Nomás llegó mi caballo y le tumbo [a] el compañero la silla. Y lo monté yo en pelo y me fui rumbo donde se fueron los caballos [de los indios.]

Como a una milla los vi parados. Me apeé de mi caballo y le puse sogue, y [mi caballo] les relinchó. Yo me acosté entre el zacate alto. Y ellos se vinieron a reconocer mi caballo. Yo tenía un mecate en la mano hecho lazo. Me fui arrastrando y me enderecé cerca de mi caballo cubriéndome con él. Quise darle para que se acercara a ellos, pero me pegó una patada cerca del estómago y me tumbó. Al extremo [ruido] que hice, [los caballos] se asustaron y se fueron corriendo como un cuarto de milla.

Yo me quedé revolcando y quejándome por un buen rato, pero nomás se me [a]minoró el dolor y monté el caballo y los eché [a los otros caballos] para el campo. Les hablé a los de mi compañía que agarraran un cabestro en dos y les eché los caballos entre la caballada de nosotros. Cuando tiraron el cabestro, el caballo se paró y la yegua voló el mecate y se fue para donde estaba la caballada de otra compañía. Aquellos soldados hicieron por agarrarlos, pero yo había montado el caballo en pelo y con bozal y seguí a la yegua.

Los soldados me dijeron: "Ésta es de nosotros", pero luego que la yegua vió al caballo se vino para donde yo estaba, y les dije: "Pues agárrenla". Y le dí galope al caballo y la yegua vino y me alcanzó y se me pegó a la pierna y le puse el mecate sin trabajo. El caballo se lo dí a los que me ayudaron a agarrarlo. Ese mismo día vendí la yegua por sesenta pesos.

CAPÍTULO 21

Ya para esa fecha se nos habían acabado las provisiones, pero como había mucho búfalo y venados, no pasábamos hambre. Pero ni señas de Buffalo Hump. En la noche [nuestros jefes] consultaron con el jefe de los indios qué harían con los indios prisioneros. La sentencia que les dieron fue cortarles la trenza y raparles la cabeza y soltarlos sin armas por embusteros. Me dijo el cautivo que era el castigo más vergonzoso para un indio. Ya estábamos en los llanos estacados, cerca de Nuevo México. De allí nos regresamos. Otro día todos los indios se vinieron para Cooper, ya de regreso.

Un día que paramos muy temprano, le pedí permiso a mi capitán de salir a matar un búfalo. Cosa de una milla del campo hallé un hatajo y herí un toro muy grande. Yo quería tener el gusto de matarlo y lo seguí. Se cortó del hatajo y se fué por la falda de unas lomas. Me quedé atrás cargando mi pistola y lo seguí. Era un terreno muy arroyento[18]. Iba yo muy violento cuando brincó el caballo un arroyo y cayó adentro, porque no pudo salvar el trecho. En la cabecera del arroyo y como a diez pies estaba un león asustado. Tal vez [se asustó] con el [hato de] búfalo que acababa de pasar y con el tropel de mi caballo. Parecía que estaba en el acto de brincar sobre mí. El caballo había caído de rodillas, y yo no me desprendí de él. Se paró muy pronto, y salí de allí temblando de miedo. Ni supe por dónde se fue el toro. Me devolví al campo y llegué ya noche. Y los compañeros estaban con cuidado de mí. Creían que los indios me habían matado.

[En] otra ocasión también le dí un tiro a [un] toro y lo seguí. Brincó un barranco y se quebró la paleta. Yo bajé al arroyo donde estaba caído y me le arrimé con la pistola en la mano. A distancia de diez pies le tiré en medio de los ojos donde estaba caído, pero nomás meneaba la cabeza a cada tiro. Esa vez me había seguído un amigo mío que se llamaba Jim Claude que era guía del gobierno y oyó los tiros. Al tirarle el último tiro, se levantó el toro y se vino sobre mí. Yo quise huir, pero me enredé con las espuelas y caí para atrás. Y el toro como tenía la paleta quebrada cayó para donde yo estaba, y con el hocico me tocó los pies. A ese tiempo llegó Jim Claude y le dió un tiro en un ojo con la carabina. Yo no sabía que tenían tanta lana en la frente. Con razón las balas no le entraban.

Después de estar en Camp Cooper como un mes, fue cuando se organizó la expedición de Wichita cuando las tropas del Mayor Van Dorn pelearon con los indios y los destruyeron. Fue una noche al amanecer — la noche del mismo día les vino la noticia a los indios de la reservación. En la noche, como a media noche, se oyó un clamor del campo de los indios, que se oyó hasta nuestro campamento a distancia de como una milla. Fuimos algunos soldados a ver y hallamos a todo el campo en pie, las mujeres cortándose los pechos con cuchillos y los hombres haciendo duelo y cantando cantos de guerra, según nos dijo un cautivo que había allí. Ese mismo día en la tarde nos vino a nosotros la noticia del pleito y [la] batalla y la muerte del teniente Van Camp. Todo esto fue durante el año 1859.

Después de la muerte de Van Camp, mandaron otro teniente segundo a nuestra compañía que se llamaba Cunningham. Éste se hizo cargo de la compañía por algún tiempo porque el capitán y el primer teniente, que se llamaba Chambers, estaban ausentes.

El día último de marzo de 1860, cumplí cinco años de servicio en el ejército de los Estados Unidos de América, y el día primero de abril de 1860 recibí mi "Honorable Discharge".

De allí me vine a San Antonio con el fin de casarme con una parienta de J. P. Rodríguez que yo había conocido en casa de él. Y había hecho trato de casamiento con ella, prometiéndole cumplir tan pronto como saliera del ejército.

A poco de haber salido yo [del ejército,] salió también el sargento de orden de mi compañía, y vino al condado de Bandera y compró cincuenta vacas con el fin de establecer un rancho de vacas. Me las dió a medias por cinco años.

CAPÍTULO 22

En ese mismo año se declaró la guerra entre los Estados Unidos y los Estados Confederados, y cuando cl cjército cmprcndió marcha para salir del estado de Texas, el sargento McDonald, que así se llamaba, se desertó y vino a donde y[o] estaba y vendió las vacas, prometiéndome $250.00 para que le entregara todas las vacas. Cuando las entregué, me dijo que viniera con él a San Antonio, donde iba a sacar el dinero en el banco. Llegamos a San Antonio y yo paré en casa de unos parientes de mi esposa, y él me dijo que otro día en la mañana fuera al hotel para pagarme. Otro día fui, y me dijeron que en la noche había montado la diligencia para Indianola, y ya no lo volví a ver. Me dejó dos caballos que él traía que valdrían $50 cada uno.

En ese año logré un terreno de 160 acres que el estado de Texas concedía a cada ciudadano cabeza de familia. Y puse un rancho en un brazo del arroyo del Privilegio, que se llamaba Bear Creek. Allí viví algún tiempo con mis cuñados,[19] que vinieron a vivir conmigo en el condado de Bandera. J. P. Rodríguez había establecido un rancho en el principal brazo del arroyo del Privilegio, como [a] dos millas distante del mío.

Los indios molestaban mucho a los habitantes de la frontera. En ese tiempo había un hombre que se llamaba John James que tenía muchos terrenos. Éste me vendió un terreno de trescientos acres,[20] cerca del pueblo de la Bandera. Yo tenía la idea de establecer un rancho allí. Este hombre había publicado en los periódicos que tenía muchos terrenos en todos los condados de la frontera de Texas y los ofrecía a peso el acre en dinero confederado. Yo le pagué $150, quedándome a pagarle el resto más allá con sus réditos. Esto fue al principio de la guerra. Como a los tres años, le fui a pagar el resto, y no quiso recibirlo, porque el dinero confederado ya no valía más que cincuenta centavos el peso. Y así quedó. Yo perdí mi dinero.

En 1862 conseguí trabajo en Campo Verde de cuidar los camellos que tenía el gobierno en ese lugar, y trabajé como ocho meses. En el otoño de ese año fui a San Antonio y entré en el ejército confederado, en el batallón del coronel Duffy.[21] Durante ese invierno, el batallón se ocupó de cuidar la línea de la parte de la frontera de Laredo a San Antonio y los puntos cercanos. Yo desde que comencé, me dieron el lugar de clarín, como ya me conocían que había servido en ese destino en ejército de Estados Unidos.

En la primavera de 1863, nos marchamos para Brownsville, y cuando llegamos a ese lugar, el batallón se organizó en regimiento con el coronel Duffy y James Sweet de teniente coronel. Durante ese año guardamos la línea del Río Grande, desde su desemboque en la mar hasta Eagle Pass. Varias veces le tocó a mi compañía guardar el lugar donde entra el Río Grande a la mar.

Por el lado de Texas estaba el estacamento de tropas confederadas, y del lado de México, un pueblito que se llama Bagdad, en el cual estaba la aduana del puerto de Matamoros. Los barcos de guerra de la marina de los Estados del Norte llegaban hasta la barra que distaba como dos millas. Y todos los días, saltaban a tierra en barcos pequeños soldados del Norte para el lado de México. También había varios barcos de vapor que entraban a Bagdad, y estaban al lado de México. En estos barcos había algunos oficiales de las tropas del Norte que salían a pasearse. Y había varios que se habían desertado de Texas, y otros que tomaron el partido del Norte y se habían pasado [al lado mexicano.] Esta clase de gente, todos los días nos provocaban gritándonos insolencias.

Del lado de Texas había un capitán que pertenecía a nuestro regimiento y que había vivido muchos años en la frontera del Río Grande. Y había sido capitán de los Rangers y conocía muy bien a los mexicanos. Y con un pico de gente, andaba vigilando la línea por el lado de Texas. Un día llegó con diez hombres a nuestro campo, y traían un mexicano prisionero. Y lo entregó a la guardia, diciéndoles que tuviesen doble cuidado de él, porque se les iba. El capitán se llamaba Brunington.

En la noche las fragatas de guerra nos tiraban algunas bombas dirigidas a las luces. Porque teníamos una casa de dos pisos y todas las noches los soldados bailaban unos con otros en el alto de la casa, y allí dirijían las bombas, pero pasaban muy altas y iban a reventar como una milla más allá. Pero esa noche del día que trajeron al hombre, nos tiraron varios tiros y nos llegaron muy cerca. Muy pronto apagamos las luces. Al preso lo tenían maniatado con una cadena de los dos pies. En la tarde del día siguiente, el capitán dió órden que se cambiara la guardia a otro lugar, porque donde estaba había peligro si seguían bombardeándonos.

El preso estaba sentado contra una trinchera de sacos de maíz, y el cabo de guardia le dijo al preso que se parara para irse al lugar donde iban a cambiarse. Pero él no quería pararse y como era tiempo de frío, estaba sentado y envuelto en su frazada. Creyendo que no entendía, me habló el cabo que le dijera que se parara, y él [me] dijo: "No, si aquí me dijo el capitán que me estuviera".

El cabo impacientado de esperarlo, lo agarró de un brazo y lo paró. Y cuando se paró se descubrió que se había quitado la cadena de los pies, y esperaba que se oscureciera para irse. Nomás se paró y se vió los pies, y dijo: "¡Mira! ¡Ya se me desató!" A esto todos los que estaban allí comenzaron a reírse y a gritar y tirar los sombreros para arriba. A los gritos vino el capitán Brunington, y dijo al preso: "¿Qué sucede, hombre?", y él con una inocencia fingida le dijo: "Se me desató la cadcna".

Entonces el capitán les dijo a los soldados: "Este mexicano se les va a ir, y al que se le vaya recibirá un castigo muy severo". Y algunos dijeron, "I'd like to see him get away from me".

CAPÍTULO 23

Esa noche como a media noche, vino de Brownsville el Coronel Chilton con un pico de gente, y a esas horas nos despertaron y nos dijeron que teníamos que pasar el Río Grande, y tomar prisioneros a ciertos individuos que había allí y que nos estaban insultando todos los días. Con mucho silencio nos preparamos, y pasamos el río en cosa de tres esquifes.[22]

El coronel Chilton nos echó un discurso en voz baja, después que nos formamos listos para pasar, en el que nos decía que no era el intento ir a derramar sangre de ningún mexicano, ni de nadie si se podía evitar, que solamente íbamos a tomar prisioneros a ciertas personas. Había ciertas personas que eran amigos nuestros, y ellos nos proporcionaron las esquifes, y habían dado santo y señas de las personas que buscábamos.

Nomás pasamos y rodeamos la casa de la aduana. El coronel Chilton y otros dos oficiales entraron y sacaron al Juez Davis, que era juez de distrito en Texas, y un gran abogado, que después de la guerra fue gobernador del estado de Texas. Y también sacaron a un hombre que se llamaba Montgomery, que según decían era malhechor.

Después los soldados supieron que en unas lomas de arena que había cerca de Bagdad, había varios Yankees y fuimos a tomarlos prisioneros. Como a un cuarto de milla del pueblo, íbamos subiendo una lomita cuando se le fue un tiro a uno de los compañeros. Ya había amanecido y al tiro vimos levantarse hombres en diferentes direcciones, y como ya era de día y estábamos en terreno mexicano, recibimos orden de devolvernos, y para esto teníamos que pasar por el pueblo, que ya estaba todo alarmado, y al acercarnos a las casas, nos recibieron con una lluvia de balas, de las casas, por las ventanas y las puertas. Pero nosotros teníamos orden de no hacer fuego. Antes de llegar al río, nos salió el juez del lugar a preguntarnos con qué fin habíamos pasado a México. Le dijimos que con el fin de llevar ciertos americanos presos, pero que teníamos orden de no atropellar la ley de México.

Uno de los compañeros que se llamaba Manuel Menchaca, nativo de San Antonio, se montó a un caballo que estaba persogado y siguió [a] dos hombres de los que estaban durmiendo en las lomas de arena y los tomó prisioneros. Él llevaba una pistola de un tiro y ellos tenían una de seis tiros cada uno.

CAPÍTULO 24

Cuando llegó [Menchaca] con sus prisioneros, le dijo al capitán: "¿Qué le parece de Menchaquita, se le dará siquiera de cabo?" Otro alemán que también se cortó de nosotros venía con dos prisioneros. Ya estando todos nosotros del lado de Texas, y nomás lo vieron [al alemán] solo, y se vinieron la turba de gente, hombres y mujeres, gritando: "¡Mátenlo! ¡Mátenlo!" Y le dieron un tiro con carabina, en la espalda, que le atravesó el cuerpo, y lo hubieran matado, pero nuestros soldados lo protegieron, formándose a la orilla del río. Y les tiraron un descargue en el que tumbaron a[l] amo que venía adelante a caballo. Y todos los demás huyeron cuando ya acabamos de pasar.

Entré yo blasfemando contra los que habían herido al alemán, al cuarto donde estaba el Judge Davis, sin saber que allí estaba. Y me dió una mirada tan respetuosa, que me avergoncé, pues era un hombre de un aspecto muy respetuoso.

Ese mismo día después de mediodía, se dió orden de llevar a los presos para Brownsville, y entre ellos iba a ir el mexicano que había agarrado Brunington. Ya cuando todo estaba listo, montaron a Montgomery en un caballo y le amarraron las manos por detrás y los pies por debajo del caballo. Y él dijo: "Todo lo que pido es que se me trate como prisionero de guerra". A lo que le contestó el coronel Chilton: "Será usted tratado como merece". Y después supe que lo habían ahorcado en el camino. A Judge Davis, le dijo Chilton: "Usted puede ir suelto en su caballo" a lo que él contestó: "Gracias".

Pero el mexicano se escapó de una manera trájica. A él y a los otros presos lo[s] iban a llevar en un carro. Ya cuando iban a salir le quitaron la cadena, y pidió permiso que lo sacara el centinela para la orilla del río, donde había unos barcos viejos que servían como de excusado. El centinela que lo cuidaba se paró como a diez pies de él. Y el [preso] con su frazada embozado, se sentó [en uno de los barcos viejos] y de repente, se zafó el pantalón y arrojó la frazada y corrió al río y se arrojó al agua.

El centinela estaba viendo para otro rumbo cuando le gritaron los que lo vieron primero: "¡Se te va el preso!", le pescó tan de sorpresa que en lugar de tirarle con la carabina que tenía en la mano, la tiró y partió tras de él sacando la pistola. El preso nomás cayó al agua y se zambulló[23] y se fue nadando por debajo [del agua] como veinticinco pies. Y sacó la cabeza, agarró resuello y se zambulló otra vez. Los soldados que tenían las armas a mano le hicieron fuego, pero ninguno le dió. Se zambulló como cuatro veces. Del otro lado del río, había varios barcos de vapor y en ellos había familias y empezaron las señoras americanas a gritar. Entonces se levantó el fuego, ya el preso estaba cerca del lado de México. Luego que cesó el fuego, bajaron una esquife y lo recibieron. Había muchos americanos [del norte] en los barcos, y empezaron a burlarse de nosotros y a gritarnos insolencias, y diciéndonos cobardes. Y siguieron insultándonos hasta que dió orden el capitán que se armaran los soldados y se formaran en la orilla del río. Y [entonces] les dijo: "Si vuelven a gritar les hacemos fuego". Los presos que nos quedaron se fueron para Brownsville.

Entre tanto por el lado de México, luego que pasamos a los presos para este lado, telegrafiaron a Matamoros. Y esa misma tarde antes de meterse el sol, llegaron dos compañías de soldados a Bagdad. El general Bee que era el que mandaba las tropas confederadas en Brownsville, tan pronto como llegaron con Judge Davis a Brownsville, lo despachó con una escolta para San Antonio.

Pero su esposa que estaba con él en Bagdad la noche de su aprehensión, tomó un coche y fue a pedir auxilio a Matamoros. Inmediatamente las autoridades militares de Matamoros demandaron del General Bee que entregara al Judge Davis, [por]que lo habían tomado prisionero de terreno mexicano. En ese tiempo como México estaba de guerra con los frances[es], había muchas tropas en Matamoros y el general Bee tuvo que ceder a la demanda del general en jefe, que me parece era Cortinas. Otro día entregaron a Davis.

CAPÍTULO 25

En ese tiempo el general Bee tuvo a bien mandar para San Antonio unos prisioneros Yankees que se habían tomado en la misma noche que se tomó al juez Davis. A mí me tocó ser uno de los que formaron la escolta. Vinimos a San Antonio, y habiendo entregado los presos, obtuve yo licencia para visitar a mi familia que estaba en mi rancho en el Condado de Bandera.

Cuando volvimos al regimiento, ya lo encontramos en Santa Gertrudis, lugar que era más bien conocido por [el nombre de] King's Ranch. El ejército de los Yankees habían desembarcado en Bagdad y el ejército Confederado se había retirado del Río Grande y marchaba rumbo a Corpus Christi.

El día que llegamos al regimiento, me rodearon mis compañeros mexicanos que habían quedado en el regimiento (pues ya muchos se habían desertado) suplicándome que hiciera una petición al coronel Duff para que nos permitiera transladarnos al regimiento del coronel Santos Benavídez, que era un regimiento de mexicanos en el servicio de los Estados Confederados. Porque durante mi ausencia había desaparecido el cabo Juan Mercado y esto fue de la noche a la mañana. Y se creía que lo habían ahorcado, pues había en el regimiento un banda compuesto de cosa de diez individuos que tenían el nombre[24] de ser ahorcadores. Y se decía que habían ahorcado a varios hombres que no estaban de acuerdo con sus ideas. Y entre ellos había un capitán que se llamaba Taylor, que decían tenía agravio con el cabo Mercado, y había hecho la amenaza de horcarlo. Y esto lo dijo delante de un amigo del cabo Mercado, que se llamaba Juan Santa Ana, y éste se lo comunicó a Mercado. Y habiendo desaparecido Mercado tan repentimamente, se creía lo habían ahorcado. Y por este motivo, los mexicanos querían ser transferidos al regimiento de Benavídez, creyendo que no tenían la vida segura los mexicanos en ese regimiento. Pues a cada paso se suscitaban disgustos entre ellos y los americanos. Pero el coronel no nos concedió la petición, diciéndonos que estábamos equivocados. Pues él nos aseguraba que Mercado no había sido ahorcado.

En ese tiempo el coronel recibió orden de marchar con el regimiento para el Estado de Lusiana. Y habiendo marchado como diez días, un día que estábamos acampados en el Río de la Vaca, se suscitó una discordia entre un mexicano y un americano. Y llegó a tanto que los americanos tomando las armas dijeron que iban a poner fin a los griseros[25] que había entre ellos. Entonces yo y un compañero mío (alemán) que había servido conmigo en el ejército del Norte por cinco años, antes de la Confederación, nos metimos de por medio y los apaciguamos. Dicho amigo mío se llama Fred Metzger y vive ahora en Hondo City, Medina County. Creo que si no hubiera sido por él, hubiera habido muchos muertos ese día.

El mismo día todos mis compañeros mexicanos se determinaron a desertarse esa misma noche, y insistieron en que yo me fuera con ellos, diciendo que nos tomaríamos los mejores caballos del regimiento y nos iríamos esa misma noche. Entonces yo les dije que si hacían eso, de robarse caballos ajenos, mucho menos me iría yo con ellos, porque si nos seguían y nos alcanzaban, nos ahorcarían por ladrones. Y que si yo los acompañaba y me agarraban, moriría por desertor, y no por ladrón. Entonces convinieron a que nos fuéramos en nuestros propios caballos.

CAPÍTULO 26

En ese mismo tiempo se supo la noticia de la toma de Vicksburg que era el punto más fuerte que tenían los Estados Confederados en el Río de Mississippi y yo comprendí que se acababa la esperanza de la Confederación. Esa noche del día primero de marzo de 1864 nos desertamos yo, Benigno Arocha, Francisco Martínez, José Garza, José Casillas y Santos Blanca. Nuestras miras eran ir a nuestras casas, visitar a nuestras familias y citar un día en cierto lugar para reunirnos y marcharnos para México.

El lugar [determinado] fue en las lomas de San Gerónimo como [a] dos millas del rancho de Don Manuel Hernández, un señor que era conocido y amigo de todos nosotros. La fecha para nuestra partida el 20 de marzo. El día citado, nos reunimos en dicho lugar yo y F. Martínez y José Casillas. Los demás no vinieron. Los esperamos varios días. A los seis días vino Santos Blanca y José Garza. Todas las noches íbamos a la casa de Don Manuel Hernández a buscar noticias. Él nos ayudaba con provisiones y nos daba noticias de lo que pasaba. Nosotros todo el día estábamos enmontados arriba de la loma, en donde había muchos cedros y desde allí veíamos a todos los que pasaban por el camino que viene de la Bandera para San Antonio.

Un día me arrojé a acercarme al camino, para ver si pasaba algún conocido, y logré ver dos que pasaban con cuatro camellos. Venían de Campo Verde, lugar donde tenía el gobierno como cincuenta camellos. Y iban para San Antonio a traer carga. Las personas eran bien conocidos míos. Era el mayordomo de los camellos, Henry Ramsey, y Candelario Treviño, que era compañero mío cuando yo trabajaba con los camellos.

Me acerqué como a diez varas de ellos, por entre el chaparral, pero no les hablé, porque yo me consideraba criminal, por ser desertor y temía que me descubrieran.

El señor Manuel Hernández todas las noches nos tocaba una bocina de cuerno de vaca que él tenía, y estábamos entendidos que si no tocaba el cuerno, era porque había alguna gente de desconfianza. Nadie de los vecinos de los otros ranchos extrañaba que tocara el cuerno, porque era la costumbre de todos los rancheros [usar un cuerno] para llamar las vacas. De esta manera nos estuvimos en ese lugar hasta el día 29 de marzo. Ese día marchamos para México yo, José Casillas, José Garza y Francisco Martínez. Ninguno de nosotros había pisado México nunca. Sabíamos que teníamos que pasar el Río Grande para llegar a México, pero no sabíamos dónde quedaba el Río Grande.

Tomamos la dirección al poniente. No quisimos tomar camino por no encontrar gente, y sabíamos que todos los días andaban partidas de soldados confederados, tanto por buscar indios, que entonces hacían mucho daño en la frontera, como también buscando desertores que todos los días iban de huída para México.

Como no conocíamos el terreno, caminábamos al tanteo el rumbo que creíamos bueno más cierto. A los tres días de ir caminando, cuando menos pensamos, llegamos a un camino muy huellado de caballada y de carros. Creímos que sería el camino que va de San Antonio para Eagle Pass. Nos paramos un rato a pensar, y determinamos tomar el camino por una corta distancia. Estábamos sobre un alto, y el camino descendía hacia una cañada.

Habíamos caminado 500 yardas cuando en una curva del camino y en el plan de la cañada y sin pensarlo, nos hallamos en medio de una compañía de soldados confederados que acababan de parar y estaban unos desensillando. Otros ponían lumbre y otros persogando sus caballos. Mis compañeros sin pararnos me preguntaron: "¿Qué hacemos?" Les dije: "Síganme".

La compañía estaba por uno y otro lado del camino. Seguimos el camino pasando por medio de ellos. Algunos de ellos se fijaron en nosotros, y otros no. Como a los 200 pasos nos paramos y les dije a los compañeros: "Esta gente tal vez nos siga. Vamos dejando el camino a la izquierda. Y nos separamos como veinte yardas los unos de los otros, para que no sigan la huella tan fácil". Lo hicimos así y como a media milla de donde pasamos los soldados, llegamos a donde estaban cuatro soldados dando agua a sus caballos, todos a pie y con las riendas en la mano, junto a un charco de agua. Ellos se fijaron bien en nosotros y, como íbamos separados, se agachaban para vernos bien por entre el monte tal vez con el fin de ver cuántos éramos.

Como una milla después que los pasamos, nos juntamos para consultar qué sería mejor hacer. Ya sería como las cuatro de la tarde. Delante de nosotros esta[ba] una lomita muy picuda y alta. El compañero Martínez llevaba el mejor caballo. Se determinó que él subiera a la cumbre de la loma, para ver si veía alguna demostración de que nos seguían, mientras los demás seguíamos el rumbo al poniente, dejando la loma como [a] una milla a la derecha. Bien vimos cuando el compañero subió a la cúspide[26] de la loma, dejándose ir para el lado del norte y dió la vuelta, y vino a donde íbamos. Nos dijo que vió algunos de los soldados, como que venían siguiendo nuestra huella. Pero que cuando lo vieron a él arriba de la loma, se dirigieron hacia él. De manera que esto nos dió tiempo a nosotros de avanzar en nuestro rumbo y con la esperanza de que ya la noche se acercaba y no podrían seguirnos la huella en lo oscuro.

Ya mi caballo iba muy cansado. Caminamos hasta media noche, y ya mi caballo no quiso dar paso. Estábamos en un terreno donde había poco monte. Les propuse a los compañeros que se fueran ellos y me dejaran a mí con mi caballo, creyendo que si otro día los soldados me alcanzaban, yo bien podría esconderme entre los chaparros. Y si hallaban mi caballo, pensarían que lo habíamos dejado cansado. Y que tal vez la siguiente noche ya podría caminar yo con mi caballo hasta llegarlo al agua. Algunos convinieron a mi proposición. Pero Francisco Martínez dijo que no, que moriríamos juntos si nos tocaba. Sin desampararnos. Entonces allí nos quedamos esa noche.

CAPÍTULO 27

Cuando amaneció empezamos a ver hatajos de caballada mesteña. Iba un hatajo como una milla distante de donde estábamos, y detrás iba un caballo cortado de los demás como [a] cincuenta yardas. Martínez dijo que ese era un caballo manso, y montó su caballo y se fue a ver si lo podía agarrar. Como era llano vimos que lo alcanzó, y luego se paró el caballo, y él se apeó y lo agarró. Y lo trajo a donde estábamos. Era un caballo viejo, del gobierno, que tal vez lo habían dejado cansado los soldados en alguna de las campañas que hacían siguiendo indios. Estaba muy flaco, pero no estaba cansado. Le eché mi silla y seguimos nuestro camino.

Ya no teníamos qué comer. Como a dos millas de donde dormimos hallamos un becerro de cosa de un mes de edad. Estaba el pobre animalito dormido entre el zacate, que había mucho. El compañero Francisco Martínez se apeó del caballo, y con un cuchillo en la mano se fue arrastrando con mucho cuidado para que no lo sintiera el becerro. Llegando a él, le sumió el cuchillo en el corazón. No pude yo menos que figurarme que Martínez era un león y el becerrito la víctima inocente.

De allí levantó cada uno la carne que quiso y seguimos con intento de detenernos para comer tan pronto como halláramos agua para nosotros y las bestias. Como a las once del día llegamos a un arroyo bastante hondo en el que hallamos algunos charcos de agua y determinamos parar dentro del arroyo y tener nuestros caballos allí, para que no se vieran, si alguna gente pasara. Había un poco de zacate y creímos que nuestros caballos tendrían bastante con eso.

Pusimos lumbre y empezamos a asar carne. Después que comimos, como ya los caballos habían trillado el poco zacate que había y ya no lo querían comer y habiendo bastante zacate fuera del arroyo, sugerí yo a los compañeros que sacáramos a los caballos fuera del arroyo para que comieran bien, y uno de nosotros se pusiera de centinela y cuidara por una hora y [los otros a] relevarlo tomando cada uno su turno. Le tocó primero a José Casillas. Y a mí me tocó relevarlo.

Me fui como doscientas varas lejos de los caballos rumbo al poniente, y [me] encontré con un camino muy huellado de carros y de caballos, como de tropa, que corría de norte a sur. Y me pensé que sería el camino que iba de Eagle Pass para Fort Clark, y yo sabía que en estos dos fortines había tropas. Al momento dí aviso a los compañeros que estábamos en peligro de que pasara alguna tropa de soldados. Determinamos meter los caballos en el arroyo otra vez, y esperar a que se hiciera más tarde para seguir nuestra marcha.

Sería como las cuatro de la tarde cuando me asomé para afuera del arroyo, y ví como diez o doce hombres con uniformes confederados que venían derecho a donde nosotros estábamos, aparentemente en nuestra huella. Vendrían como [a] 300 varas.

Al momento tomamos nuestras armas, y nos preparamos para pelear. Considerando que ellos no nos veían, nosotros teníamos la ventaja. No había tiempo que perder mas dejarles que llegaran muy cerca, y darles un descargue, si posible era matar cada uno el suyo y seguir peleando hasta morir o vencer. Estábamos en una posición que nosotros los podíamos ver, y ellos no podían vernos hasta llegar muy cerca de nosotros. Ya vendrían como a cincuenta yardas, cuando de repente, el que venía adelante, que creo sería el jefe, se paró. Habló con los demás. Y de allí cambió de rumbo y se fueron directamente al norte. Nosotros nos quedamos viéndolos hasta que se perdieron de vista. Este parece haber sido un hecho Providencial.

CAPÍTULO 28

Serían como a las cinco de la tarde cuando salimos de allí y caminamos rumbo al poniente. Como a las doce de la noche, comenzamos a descender una cuesta abajo muy larga, y yo les dije a mis compañeros que creía que ya íbamos llegando al Río Grande. A poco llegamos a un carrizal por una vereda muy honda y angosta. Yo iba delante, mi caballo cansado[27] tras de mí, y los compañeros siguiéndome.

Caminamos por aquella vereda por entre el carrizal y el monte, sin poder ver más que las estrellas. Al último llegamos a la playa del río, y me paré y les dije a mis compañeros: "Éste es el Río Grande. Ahora, ¿cómo lo vamos a pasar?" Algunos de ellos dudaban de que fuera el Río Grande. Yo les dije: "Éste es, y nomás que lo pasemos, voy a gritar: ¡Qué viva México!"

Nos acercamos a la orilla del agua y les dije: "Espérenme aquí, voy a pulsar[28] el río". Me quité toda mi ropa y agarré un palo en la mano como bordón. El agua estaba muy fría, pues era en el mes de febrero. La corriente me quería llevar, pues estaba bastante fuerte, pero yo batallé hasta que salí a una isla [en medio del río] la que atravesé y llegué al otro brazo del río. Entonces me vino la idea de que mis compañeros no estaban seguro[s], pensando que podía pasar alguna ronda[29] de soldados de los que vigilaban la línea del río día y noche. Vine y los pasé a la isla. Y me fui a sondear el otro brazo [del río.] Estaba lo mismo de hondo [y] la corriente lo mismo, pero salí al otro lado contra de un barranco muy alto. Y como era de noche, no se veía el fin de él. Seguí la orilla del agua por contra el barranco como cien varas, y hallé una vereda muy incómoda, pero evidentemente que era la única salida. Me regresé y traje a mis compañeros. Nos apeamos de los caballos y subimos cabestreándolos.

Nomás subió el último compañero, y grité: "¡Qué viva México!" y en contestación, empezaron a ladrar unos perros. No muy lejos, nos acostamos a dormir como si estuviéramos en casa.

Cuando amaneció, descubrimos que estábamos en una labor, y [había] unas casas como [a] 400 varas de donde estábamos. Les dije a los compañeros que me esperaran en lo que yo iba a buscar qué comer. La gente estaba levantándose. Les saludé y les dije que éramos desertores de los Confederados y [que] queríamos comprar algo para comer. Me preguntó el dueño de la casa si alguno de nosotros era de[l] condado de la Bandera. Yo le contesté que ese era el lugar donde yo vivía. Me preguntó si conocía a Don Andrés Gonzales. Le dije: "Lo conozco muy bien, a él y a su familia". Y le dí razón de todos por nombre, pues conocía bien a todos. Ese señor nos recibió con mucho cariño.

Después del almuerzo fuimos a presentarnos con el alcalde de Monclova Viejo. Este señor salió ser pariente de mi esposa. Y nos recibió muy bien y nos dió un pasaporte para que pasáramos a Piedras Negras. Nos aconsejó que dejáramos allí nuestras armas, pues como México estaba en guerra con los frances[es], no permitían gente armada que fuera desconocida[30]. Después vinimos y las llevamos.

Después de estar en Piedras Negras un tiempo, mandé llevar mi familia, y me fui a vivir al Presidio del Río Grande, que ahora se llama Guerrero de Coahuila.

Allí encontré a la familia del cabo Juan Mercado, que creíamos había sido ahorcado por la gavilla del capitán Taylor. Se había desertado, y había mandado por su familia, que estaba en el río de Medina, Texas. Y después de estar con su familia un mes, se suicidó. Dicen que perdió el juicio, y, en estado de demencia, se quitó la vida.

Yo viví un año en este pueblo del Río Grande, y durante este año encontré un doctor americano que se llamaba Cooley.[31] Y este me solicitó que fuera con él como intérprete a visitar las villas del estado de Coahuila. Me ofreció pagar un peso y la comida por día, para mí y mi caballo. Anduvimos varios pueblos y no hallamos enfermos. Un día me dijo: "Si yo hallara un mexicano rico, con una calenturita intermitente, le sacaba todos nuestros gastos. Le daría una dosis de medicina que lo hiciera sentirse tan malo, que pensara que se iba a morir. Y lo tendría así unos días hasta que le sacara bastante dinero para nuestros gastos".

El día siguiente llegamos a la villa de Nava, y allí había un señor rico que se llama[ba] don Felipe de la Garza. Un señor yerno de él nos dijo que su suegro esta[ba] enfermo y que desearía que el doctor lo fuera a ver. Y él mismo nos llevó. Don Felipe de la Garza era un señor anciano, muy blanco y muy gordo. Nadie podría decir que estaba enfermo. Estaba sentado sobre una cama muy alta[32] y muy ricamente adornada. Su esposa estaba en otra cama en la contra esquina de la misma sala. La señora también era muy blanca y gorda. Ambos parecían ser de la misma edad, como de 65 o 70 años.

El señor Fulgencio de la Garza (que así se llamaba el yerno) le presentó al doctor la enfermedad de don Felipe. La tenía en las piernas. Hacía como quince años que no podía andar. Varios doctores lo habían tratado, pero no lo pudieron sanar. El Dr. Cooley lo examinó y le dijo que él creía [poder] sanarlo. Si él quería ponerse en sus manos, le propuso que le compondría unas medicinas que le duraran tres semanas o un mes, y [que] volvería a verlo a fin de ese tiempo. A lo que don Felipe convino. Pero su esposa, que nomás estaba oyendo, le dijo: "Pregúntale cuánto te cobra, no sea que te quiera sumir un ojo y te saque el otro". El doctor contestó que por cinco pesos le componía las medicinas que le durarían por un mes, a lo que la señora le contestó: "¿No te lo dije? Dile que no, que te venda dos reales de la medicina, y si te alivias le comprarás mas. Si no, no".

Con aquello nos despedimos de ellos. Tan pronto como estuvimos solos, le dije yo al doctor, "¿Qué le parece de los mexicanos ricos, doctor?" Me contestó que eran malas esperanzas. Anduvimos un mes recorriendo las villas del estado de Coahuila y no encontramos más que una mujer enferma, en la villa de Zaragoza. Nos hospedamos en casa de don Anastacio Gómez, cuya esposa era tía de mi esposa. La señora enferma era esposa de un hijo de don Anastacio Gómez. Estuvimos allí cuatro días. Nos dieron buena asistencia para nosotros y nuestros caballos, y no nos quisieron cobrar nada, esperanzados que el doctor no les cobraría por la cura de la mujer. Pero el doctor les cobró nueve pesos. Ellos se quejaron que era mucho, pero el doctor les dijo que si no hubiera sido por la asistencia que nos habían dado, les hubiera cobrado cincuenta pesos. Ése fue el único dinero que ganó en todo el viaje. Nos regresamos a Guerrero y no volví a ver al doctor.

CAPÍTULO 29

En ese tiempo había dos americanos que vivían en un rancho que habían rentado como [a] tres millas del pueblo. Estos hombres se habían ido de Texas huyendo de la guerra, y habían sembrado la labor de ese rancho. El rancho quedaba en el camino que va del Río Grande a Nava, y como las tropas mexicanas pasaban por allí, y ellos no hablaban el español, las tropas del partido Imperial pasaron y les despojaron de todo lo que tenían. Y no les dejaron ni en qué dormir ni qué comer. Al mes pasaron otros y les volvieron a quitar todo lo que tenían. Uno de ellos se desesperó y se fue. El otro vino y me convidó que me fuera a vivir en el rancho con él, dándome la parte del otro, que era la mitad de la cosecha. Ya casi todo el trabajo estaba hecho. Me fui con él. Ya era el mes de junio de 1865.

En el mes de agosto, Freeman, que así se llamaba el americano, como supo que ya se había acabado la guerra,[33] me suplicó que le prestara mi caballo para venirse a ver a su madre que vivía en Halletsville, Texas, diciéndome que si no volvía, me pagara mi caballo con la cosecha, lo que sucedió. Para fines del mes de octubre, acabé yo de levantar la cosecha, y me vine con mi familia para Texas. Pasé el Río Grande en Piedras Negras (que ahora se llama ciudad Porfirio Díaz.) Yo venía con mi familia en un carro con una yunta de bueyes.

A los cuatro días de haber caminado, paramos a sestiar en Río Frío, cerca de un rancho de un americano que se llamaba Griner. A poco de haber llegado nosotros allí, llegó un negrito arreando unos caballos, entre los que llevaba, una yegua y un muleto de tres años y caballos míos que los indios se habían robado de mi rancho hacía dos años. Le pregunté al negrito de dónde habían comprado esas bestias. Y él me dijo que los indios las habían dejado cansadas y se habían llevado los caballos del señor Griner. Fui yo y hablé con él, y salió ser conocido mío, pues había sido soldado en la misma compañía donde yo estaba, en el regimiento de Duff, del cual me había desertado. Luego que me conoció, me entregó mis bestias sin vacilar. Esta buena suerte tuve en el camino, pues ya nunca pensaba haber visto mis bestias [otra vez].

Llegué a mi rancho a fines del mes de octubre. Y como mis cuñados estaban viviendo allí, dejé mi familia, y fui a buscar trabajo en Campo Verde, en donde estaban los camellos del gobierno. En ese tiempo estaban a cargo de mi amigo Fred Metzger, y estando necesitando un hombre, al momento me dió trabajo. Estuve allí hasta junio 1866. En ese tiempo el gobierno vendió los camellos, y yo me contraté con una señora viuda, a cuidar un ganado de vacas, por la mitad de las crías. En el año 1867 en octubre hice un trato con mi amigo Metzger de vivir en un rancho que compró en North Creek cerca de la villa de Comfort [y] cerca del Río de Guadalupe, en el condado de Kerr.

Ese año había yo puesto un ganado de cabras, de cosa de mil cabezas. Mi suegra vivía con nosotros. En el mes de noviembre ella se enfermó, y murió. Mi cuñado, Blas Torres, había ido a verla y él y yo, y mi esposa, la vimos morir. Murió antes de amanecer el día 25 de noviembre. Ese día, en lo que mi cuñado hizo el ataúd, yo fui a Comfort, cosa de cinco millas de allí, y conseguí un carruaje de dos caballos. Y salimos del rancho con el cuerpo, para el Privilegio, distancia de veinte millas, a las 7 p.m.

El rancho quedaba entre unas lomas muy altas, y la luna se veía por sobre la cumbre de las lomas, tan brillante y tan hermosa, que nunca se me olvidará el tiempo de la llena de la luna. Mi esposa y yo, y una niña que teníamos de seis años, veníamos en el mismo carruaje. Mi cuñado venía delante a caballo. En ese tiempo los indios atravesaban esa parte de la frontera de Texas, robando caballos y matando a toda la gente que encontraban. Y esto lo hacían cada mes cuando la luna estaba llena.[34]

Ésa fue una de las noches mas tristes que he pasado en mi vida. Tuvimos que atravesar el Río de Guadalupe y caminar por un camino muy escabroso y lóbrego. Al salir del rancho con el cuerpo, nos habíamos encomendado a Dios, y Él nos acompañó, ¡bendito sea Su Santo Nombre! Llegamos al rancho donde vivía mi cuñado, en el condado de Bandera, como a las cinco de la mañana. Ya venía amaneciendo. Su esposa estaba sola esperándolo. Cuando oyó el ruido del carruaje, se levantó y salió, preguntando por mi suegra, porque sabía que estaba enferma. Se dirigió al carruaje, y al ver a mi esposa y el ataúd, comprendió lo que pasaba, y abrazó a mi esposa llorando. Después de haber sepultado a mi suegra, regresamos yo y mi esposa al rancho de North Creek.

En el mes de febrero llegó el tiempo de la enfermedad de mi esposa, que estaba para dar a luz [a] una criatura. Le supliqué a mi cuñado que dejara a su esposa ir a estar con la mía, durante el período de su enfermedad. Pero él se negó, y tuve que traer a mi esposa a su casa, que era un jacal muy mal construido, donde el frío penetraba sin poderlo evitar. En ese tiempo se vino una temporada de nieve y [un] norte muy frío que duró como quince días. Y como la casa estaba tan mal abrigada, mi esposa agarró frío y murió. No pudo el doctor libertarla. Una señora americana asistió a mi esposa. Si hubiera estado en mi casa, creo que no hubiera muerto, pues tenía muy buena casa. Me quedaron tres niñas y dos niños.

Esto fue el año de 1868 en febrero. Ese año me vine a vivir al rancho de don Miguel Indo, en el Río de San Antonio, veinticinco millas abajo de la ciudad. Sembré ese año allí, y después que acabamos de sembrar, me puse a viajar con un tren de carretas al puerto de Powderhorn y Indianola. Esto fue antes que los caminos de fierro llegaran a San Antonio.

En el mes de mayo [de 1869] encontré en San Antonio a la señora Anastacia, viuda de Mercado. Y en el mes de junio, me casé con ella. Tenía un niño y una niña de familia. [Después de la boda] nos vinimos a vivir al rancho de don Miguel Indo donde yo tenía mi siembra.

El día 4 de julio se vino una creciente en el Río de San Antonio, y inundó toda la labor, y se perdieron todas las cosechas. Después de la inundación, resultaron una[s] enfermedad[es] de fiebre malaria en todo el país donde alcanzó la inundación. Yo y mi familia nos enfermamos. Y mi madre política y otros parientes de mi esposa vinieron y nos llevaron. Yo estaba tan mal que no supe cuándo me sacaron de la casa con todo y familia, hasta en la noche que se me minoró la calentura. Esa noche dormimos en el campo, y otro día llegamos a Medina, al rancho de los Herreras, que se llama El Paso de los Garzas. Allí estuve enfermo como un mes sin poder trabajar.

En esos días supe que buscaban un hombre para que fuera para México con un judío comerciante que se llamaba Gold Frank. Frank iba en un buggy con un tiro de mulas y llevaba dos hombres, un[o] a caballo y otro de cochero. Un día iba uno a caballo y otro le arreaba el buggy, y así nos relevábamos. Yo como iba enfermo de fríos y calenturas, el día que me tocaba ir a caballo sufría mucho. Y el día que arreaba el buggy, descansaba. El negocio de Gold Frank en México era recoger dinero de los comerciantes que le debían, pues el los surtía de efectos de San Antonio, y cada año hacía viaje a traer dinero. Pasamos el Río Grande en Eagle Pass y fuimos a Zaragoza, Morelos, Nava, Allende, Santa Rosa y Monclova, y de allí regresamos.

Después que llegamos a San Antonio, me determiné a irme para mi rancho, en el Privilegio, Condado de Bandera. Esto fue el año de 1869 en noviembre. En el año de 1870, sembré en mi rancho y levanté buena cosecha de maíz. El año siguiente me fui a vivir al rancho de John James cerca de Bandera City, como a milla y media del pueblo. Allí viví dos años. En 1874 me fui a vivir a mi rancho. En ese año, J. P. Rodríguez era Juez de Paz del precinto[35] del Privilegio, como le decíamos [a esa comunidad]. Ese año me metí a varillero. Compré mercancías en San Antonio, y saqué licencia para comerciar en los condados de Bandera y Kerr. Vendía toda clase de mercancías, de abarrote y de ropa. Y también compraba cueros de res y de toda clase de pieles. También vendía licor, pero lo hacia de contrabando. Tenía ciertos marchantes que me lo encargaban, y me daban cueros muy baratos, porque eran robados.

CAPÍTULO 30

Había entonces mucho ganado mesteño en la frontera, y yo hacía mis viajes hasta más arriba de la cabecera de los ríos de Medina y Guadalupe, en donde no había habitantes [permanentes.] Pero había campos de americanos que se [a]campaban por semanas y meses cazando venado, y osos, y guajolotes, y buscando panales de abejas. Pero mataban más vacas y novillos y toros y becerros, con el fin de vender los cueros. Yo iba a estos campos acompañado de un pariente de mi esposa, que se llamaba Jacinto Treviño.

En uno de estos viajes encontramos [a] cuatro hombres americanos con un carro tirado por una yunta de bueyes, cargado de cueros frescos y secos, y también carne fresca y seca. Venían bajando una loma muy incómoda y pedregosa. No había camino, nomás las huellas de su mismo carro, que nosotros íbamos siguiendo.

Luego que nos vieron echaron mano a sus armas y nos preguntaron qué negocio nos llevaba a ese lugar. Yo les dije que andaba vendiendo efectos[2] y comprando cueros, que habíamos sabido que había un campo de cazadores y que los íbamos buscando para venderles algo y pa[ra] comprarles cueros, si los querían vender. Ellos me contestaron que todos los cueros que ellos tenían eran robados porque ellos mataban animales de todos fierros, creyendo que los dueños de esos animales nunca los buscarían, siendo que [los dueños] vivían en donde el país estaba lleno de habitantes y sin peligro, "y nosotros vivimos aquí en la frontera, expuestas nuestras vidas y las de nuestras familias, a ser asesinados por los indios".

Todo esto nos dijeron en un espíritu amenazante, y con las armas en la mano. Yo tuve mucho cuidado de no contrariarles en nada. Les contesté que yo lo que quería era vender mis efectos y comprar sus cueros. Pero ellos me dijeron que tenían desconfianza de nosotros. Yo les aseguré que no corrían ningún peligro, que quería que fueran mis marchantes, y yo [quería] ser su amigo. Con ese fin saqué una botella de whiskey y les ofrecí un trago, con lo que se alegraron un poco y nos invitaron [a] que fuéramos a sus casas. Vivían cerca de las cabeceras del Río de Medina. Fuimos con ellos y cuando llegamos, nos compraron un galón de whiskey. Y comenzaron a beber y a comprarnos efectos en cambio de cueros. Nosotros nos [a]campamos como cien yardas de sus casas. Allí trajeron [a] sus familias y compraron indianas, imperial, y de todos los efectos que traíamos. Nos vendieron como cien pesos de cueros, y de miel de abeja, y de manteca de res. Y nos compraron más whiskey y nos encargaron muchas cosas para nuestra vuelta. Nomás cerró la noche y comenzaron a tocar el violín, y a bailar y a gritar. Como a las 10 o a las 11 p.m. vino uno de los más viejos a nuestro campo, y nos dijo: "Amigos, es bueno que ustedes se vayan, porque los muchachos están borrachos [y] quieren venir a matarlos". A esas horas prendimos los caballos [al carro], y nos fuimos. Y ya no los volvimos a ver, porque tuvimos miedo volver.

En otra ocasión fuimos a buscar cueros a unos ranchos de Bandera para abajo, en el Río de Medina. Llegamos a un rancho que estaba solo, arriba de una loma alta. Las puertas de la casa estaban todas cerradas, pero se conocía que vivía gente en ella. Salía una huella de carro fresca de allí rumbo hacia unas lomas de cedro que se veían como a tres o cuatro millas distante. Mi compañero se ocupó un buen rato en examinar los alrededores del rancho, y como a cincuenta yardas de la casa, se halló un montón de cueros entre un chaparral. Y vino y me avisó. Y entonces nos determinamos a seguir las huellas del carro, con el fin de hallar la gente y comprarles los cueros. En efecto, montamos nuestro carro y fuimos como cuatro millas entre las lomas cedrosas. El día esta[ba] frío y empezábamos a querer poner lumbre, cuando vimos un humo como a una milla distante en el rumbo que llevábamos, y nos determinamos llegar a donde estaba.

Íbamos como [a] cien yardas cuando vimos a un hombre, y una mujer, y unos niños. Le dije a mi compañero: "Estése aquí. Déjeme ir a ver qué están haciendo". Luego que me vió el hombre agarró su carabina en la mano, y se vino a encontrarme. Y como a veinte yardas me dijo:

"¿Qué busca usted aquí?" Le contesté [que yo] creía que venía buscándolo a él, porque había estado en su rancho, y había visto la huella de su carro y la habíamos seguído con el fin de venderle algunos efectos. Me preguntó qué clase de efectos traía, y le invité a verlos. Me acompañó al carro, y lo primero que hice fue ofrecerle un trago de whiskey. Lo tomó, y nos invitó a que pasáramos para su campo a calentarnos.

Cuando llegamos, la señora parecía tener desconfianza de nosotros, pero no dijo nada. Pero el hombre me dijo: "Ésta es la manera que hago mi vida. Hay mucho ganado en estas lomas, que pertenece a gentes que viven muy lejos, en lugares donde el país ya está habitado. Y nunca volverán a verlos, y nosotros estamos aquí expuestos a que nos maten los indios. Yo no puedo dejar mi familia sola en la casa y la saco conmigo cuando salgo a matar algunas vacas o novillos, para vender los cueros y ayudarme a vivir".

En fin, nos hicimos amigos, y le traté de comprar sus cueros. Y me dijo: "No tengo más que estos cuantos". Tenía como seis novillos muertos, en un trecho como de trescientas yardas, que los había matado a balazo esa mañana, y estaba quitándoles los cueros. Para cuando nos separamos de él, ya nos habíamos hecho amigos de confianza y le compramos los cueros que tenía allí y también los que tenía escondidos cerca de la casa. De allí en adelante fue uno de los mejores marchantes que tubimos.

Durante el tiempo que andaba comprando cueros y comerciando, yo me envicié en tomar licor a tal grado que ya me dominaba el vicio. Y al mismo tiempo, iba perdiendo el respeto a lo ajeno. Les compraba cueros robados a todos los vecinos del Privilegio que se atrevían a matar animales ajenos, y lo mismo lo hacía con todos, americanos o mexicanos. Les decía: "Tráigame los cueros, y no me importa de qué fierro son, aunque sean de mis vacas". Y esto los animaba a matar los animales ajenos. Algunos me traían los cueros a la casa de noche, y me los vendían muy baratos. Un cuero que valía tres o cuatro pesos, me lo daban por una botella de vino y 50 centavos de café, y me había captado la amistad de todos los malvados y estaba bien quisto con todos.[36]

Los cueros los llevaba a San Antonio cada vez que completaba viaje de cincuenta hasta cien cueros. Tenía que inspectarlos antes de sacarlos del condado de Bandera. Pero yo tenía comprado al inspector. Después de pagarle diez centavos por cada cuero, le regalaba una botella de whiskey y conseguía con él que pusiera en la lista de inspección los fierros que yo le decía, aunque no los tuvieran los cueros. Todo eso y mucho más se puede hacer con el licor. El inspector se conformaba con decir "Quiero vivir y dejar vivir", sin oír la voz de la conciencia, ni yo, ni él. Yo para tener mas contentos a mis vecinos, de cuando en cuando les hacía baile en mi casa.

En ese tiempo, era Juez de Paz del precinto J. P. Rodríguez. Pero nadie de los vecinos lo querían, porque él quería prohibir el mal de matar vacas ajenas y vender cueros robados. Y aunque él andaba con nosotros en los bailes y carreras de caballos, y demás desórdenes, [todos] tenían que cuidarse de él, para matar animales ajenos y vender los cueros.

En el ano 1875, se llegó el tiempo de las elecciones. Y mis amigos me postularon a mí para Juez de Paz del precinto. Y como yo era malvado, me querían para juez. Y el día de la elección, el Señor J. P. Rodríguez no sacó ni un solo voto.

Después de las elecciones, a su debido tiempo, me llamaron al pueblo de la Bandera, para que tomara el juramento como Juez del precinto. Y cuando me dijeron: "¿Jura Ud. obedecer y hacer que se obedezcan las leyes del Estado de Texas?"

NOTAS

[1] El pueblo a cual se refiere se llamaba Council Grove.

[2] Goletas. Posible influencia del inglés (schooners).

[3] No se sabe que significaba la palabra "majoros" para Santiago, pero es posible que fuera un arcaismo o coloquialismo y una variación fonética de la palabra "masoros", para describir los que estudiaban la Masorah, la obra Hebrea que se usa para preservar la Biblia de corrupción de contenido. Este término, posiblemente vino a significar los que estudian los textos religiosos. Otra posibilidad es que se refería a los "masones", aunque no se sabe si los masones existían públicamente en Nuevo México en ese tiempo, ya que la iglesia católica se oponía a las sociedades masónicas a través de las declaraciones papales en contra de masones de 1738 a 1917, (y más recientemente, declaraciones antimasónicas hechas por el Cardenal Ratzinger, ahora Papa Benedicto.)

[4] Se refiere a Talbotton, Georgia, la sede del Instituto Metodista que él llama Talbaton Institute.

[5] Sic. El original dice "hablaban of the Mexican boy." Santiago frecuentemente cambia de idioma en su relato, a veces usando la palabra en inglés para hacer puente a la narración en español.

[6] Un "Norther" es palabra coloquial texana para describir un viento del norte que llega de pronto y de un frío extremo; es típico de las temporadas más frías en Tejas.

[7] Santiago usa la palabra campo en varios lugares para decir campamento.

[8] En referencias siguientes este nombre está deletreado "Prusa" por Santiago.

[9] Fidel interpreta "paramos" como acampar.

[10] "Hello!"

[11]"Porque también yo soy hombre bajo de potestad, y tengo bajo de mí soldados: y digo a este: Ve, y va; y al otro: Ven, y viene; y a mi siervo: Haz esto, y lo hace". *La Santa Biblia*, Casiodoro de Reina/Valera, N. Y. Capítulo 8: Versículo 9.

[12]Palabra coloquial que se usa para describir las gotas heladas cuando está granjeando, o lluvia helada.

[13]La jugada.

[14]Kickapoo Springs en inglés.

[15]"Bofes", arcaismo para pulmones.

[16]traspasó las mangas.

[17]No podían llevar su caballo al trote.

[18]De muchos arroyos.

[19]Santiago lo deletrea "cuyados".

[20]En el original, Santiago parece haber escrito el número "2" arriba del "0", indicando que eran 320 acres.

[21]El apellido correcto del oficial era "Duff".

[22]esquifas.

[23]sumergió.

[24]la mala reputación.

[25]Traducción al español de la palabra "greasers", palabra despectativa para mexicanos. Hay una ley, "The Anti-Vagrancy Law" que hace mención de este término.

[26]la cumbre.

[27]El original dice "casado".

[28]a sondear.

[29]alguna patrulla.

[30]No permitían a los extranjeros que cargaran armas.

[31]El original dice Coley, pero Fidel dice que Santiago lo había pronunciado "Cooley".

[32]En el original está escrito "alba" pero Fidel transcribe "alta".

[33]La guerra civil en los Estados Unidos.

[34]El original dice "cuando la luna estaba buena".

[35]distrito electoral.

[36]Y mantenía amistad con todos.

WORKS CITED

"A Sesquicentennial Timeline: 1833-1987." 13 Aug. 2008. *Emory University.* <http://emoryhistory.emory.edu/timeline/index.html>.

Banks, Stanley C. "The Mormon Migration Into Texas." *Southwest Historical Quarterly* October 1945: 49.

Bartlett, John Russell. *A Personal Narrative of Explorations and Incidents in Texas, New Mexico, California, Sonora and Chihuahua 1850-1853*. Vol. II, Appendix C. 383-384, 400-402. Print.

Chalfant, William Y. *Without Quarter: The Wichita Expedition and the Fight on Crooked Creek.* London: U of Oklahoma P, 1991. Print.

"Columbus Muscogee County." *Georgia.gov* 22 March.2009 <http://columbus-muscogeecounty.georgia.gov/>.

Davis, Charles G. "Camp Cooper." 15 Jan. 2008. *The Handbook of Texas Online.* <http://www.tshaonline.org/handbook/online/articles/CC/qbc9_print.html>.

Drummond, Loren. "St. Louis County Parks & Recreation: Jefferson Barracks Historic Park." Jun. 2002 <http://www.civilwaralbum.com/misc/jeff barracks1.htm>.

"Don José Manuel Gallegos." 29 Oct. 2007. *The Library of Congress.* <http://lc web2.loc.gov/ammem/today/oct30.html>.

Emmett, Chris. *Texas Camel Tales.* San Antonio: Naylor, 1932. Print.

Foley, Neil. *White Scourge: Mexicans, Blacks and Poor Whites in Texas Cotton Culture.* Berkeley: U of California P, 1998.

"Franklin Pierce 1853-1857." *The White House.* <www.whitehouse.gov/about/presidents/FranklinPierce/>.

"Fort Belknap." *Graham.* <http://visitgrahamtexas.com/Attractions/Detail/?ID=18>.

"Galisteo, NM." Map. *MapQuest.* 2009. MapQuest.com, Inc. 9 July 2009 <http://www.mapquest.com/maps?city=Galisteo&state=NM>.

García, Marina, and Kyle Putnam. "Corridos y Tragedias de la Frontera." 5 Jun. 2001 <http://www.laits.utexas.edu/jaime/jnicolopulos/cwp3/icg/index.html>.

"General Winfield Scott and the Mexican-American War 1846-1848." *About North Georgia.* <http://ngeorgia.com/other/scottinmexico.html>.

"Historic Camp Verde." 23 Jun. 2007. *Kerr County Historical Commission.* <http://www.co.kerr.tx.us/historical/camp_verde.htm>.

Hunter, J. Marvin. *Old Camp Verde: The Home of the Camels: A Romantic Story of Jefferson Davis' Plan to Use Camels on the Texas Frontier.* Bandera, Texas: Hunter, 1948. Print.

"Jefferson Davis 1808-1889 President of the Confederate States of America." <http://americancivilwar.com/south/jeffdavi.html>.

Johnston, William Preston. *The Life Of General Albert Sidney Johnston: Embracing His Services In The Armies Of The United States, The Republic Of Texas, And The Confederate States.* New York: Da Capo P, 1997. Print.

Kendall, George Wilkins. *Narrative of the Texan Santa Fe Expedition. Chicago*: The Lakeside P, 1929. Print.

"Las Vegas, NM." Map. *MapQuest.* 2009. MapQuest.com, Inc. 9 July 2009 <http://www.mapquest.com/maps?city=Las+Vegas&state=NM>.

Marroquín B. Guerra, Herlinda. Personal interview conducted by Carmen Tafolla. 3 April 2009.

"Menger Hotel San Antonio, Texas." *National Trust for Historic Preservation.* <http://www.historichotels.org/hotel/the_Menger_Hotel/>.

"The National Hotel." *Mr. Lincoln's White House.* <www.mrlincolnswhitehouse.org/inside.asp?ID=185&subjectID=4>.

Olmsted, Frederick L. *A Journey Through Texas*. Austin & London: U of Texas P, 1978.

"Pecos and Staked Plains, Physiographic Area 55." *Bureau of Land Management.* <www.blm.gov/wildlife/pl_55sum.htm>.

"Presidio Military Records Dec. 27, 1837" *State Records Center*. Santa Fe, New Mexico, accessed 1970.

"Preston, TX." Map. *MapQuest.* 2009. MapQuest.com, Inc. 10 July 2009 <http://www.mapquest.com/maps/map.adp?city=Preston&state=TX&country=us>.

"Red River Mississippi Watershed." 5 Jun. 2009. *Wikipedia.* <http://en.wikipedia.org/wiki/Red_River_(Mississippi_watershed)>.

Riley, John Denny. "Santos Benavides: His Influence on the Lower Rio Grande, 1823-1891." Diss. Texas Christian University, 1976. Print.

Rodriguez, Jose Policarpo. *The Old Guide: Surveyor, Scout, Hunter, Indian Fighter, Ranchman, Preacher.* Nashville & Dallas: Publishing House of the Methodist Episcopal Church South, 1898. Print.

"Santa Fe National Historic Trail." 20 Jun. 2007. *U.S. National Park Service.* <www.nps.gov/safe/historyculture/index.htm>.

Santos, Richard. *San Antonio de Bexar—1 de Enero 1836.* San Antonio, 1963. Print.

Sowell, A. J. *Early Settlers and Indian Fighters of Texas*. Austin & London: U of Texas P, 1900. Print.

Tafolla, Esther. Personal interview conducted by Carmen Tafolla. 19 June 1971.

Tafolla, Mary D. Personal interview conducted by Carmen Tafolla. 2 January 2009.

Tafolla, José Mariano, "Letter from Sevilleta, 22 Feb. 1816." State Archives of New Mexico II, Collection #1959-200, Twitchell #2650. New Mexico State Records Center and Archives.

Tafolla, Mariano Jr. Personal interview conducted by Carmen Tafolla. October 20 1972.

_____. Personal interview conducted by Carmen Tafolla. December 22 1972.

_____. Personal interview conducted by Carmen Tafolla. December 27 1972.

Tafolla, Sarah. Personal interview conducted by Carmen Tafolla. July 18 1964.

_____. Personal interview conducted by Carmen Tafolla. August 2 1967.

_____. Personal interview conducted by Carmen Tafolla. June 21 1972.

Thompson, Jerry Don. "A Stand along the Border: Santos Benavides and the Battle for Laredo." *Civil War Times Illustrated.* Aug. 1980. Print.

"Talbot County." *Georgia.gov* <http://talbotcounty.georgia.gov>.

Taylor, T. U. "Lyman Wight and the Mormons in Texas." *Frontier Times.* June 1941.

Thompson, Jerry. "Juan Nepomeceno Cortina." 17 Jan. 2008. The Handbook of Texas Online. <http://www.tshaonline.org/handbook/online/articles/CC/fco73.html>.

"Timeline of Texas and the Western Frontier 1856-1865." 30 Jun. 2003. *Texas Beyond History.* <http://www.texasbeyondhistory.net/forts/56-65.html>.

"Treaty of Guadalupe Hidalgo." 23 Jun. 2009. *Wikipedia.* <http://en.wikipedia.org/wiki/Treaty_of_Guadalupe_Hidalgo>.

Twitchell, Ralph Emerson. *The Leading Facts of New Mexican History Vol.2.* Cedar Rapids, Iowa: The Torch P, 1912. Print.

Vestal, Stanley. *Wagons Southwest—Story of Old Trail to Santa Fe*. New York: American Pioneer Trails Association, 1946. Print.

Wight, Levi Lamoni. *The Reminiscences and Civil War Letters of Levi Lamoni Wight.* Salt Lake City: U of Utah P, 1970. Print.

Williams, R. H. *With the Border Ruffians: Memories of the Far West, 1852–1868.* New York: Dutton, 1907. Print.

APPENDICES

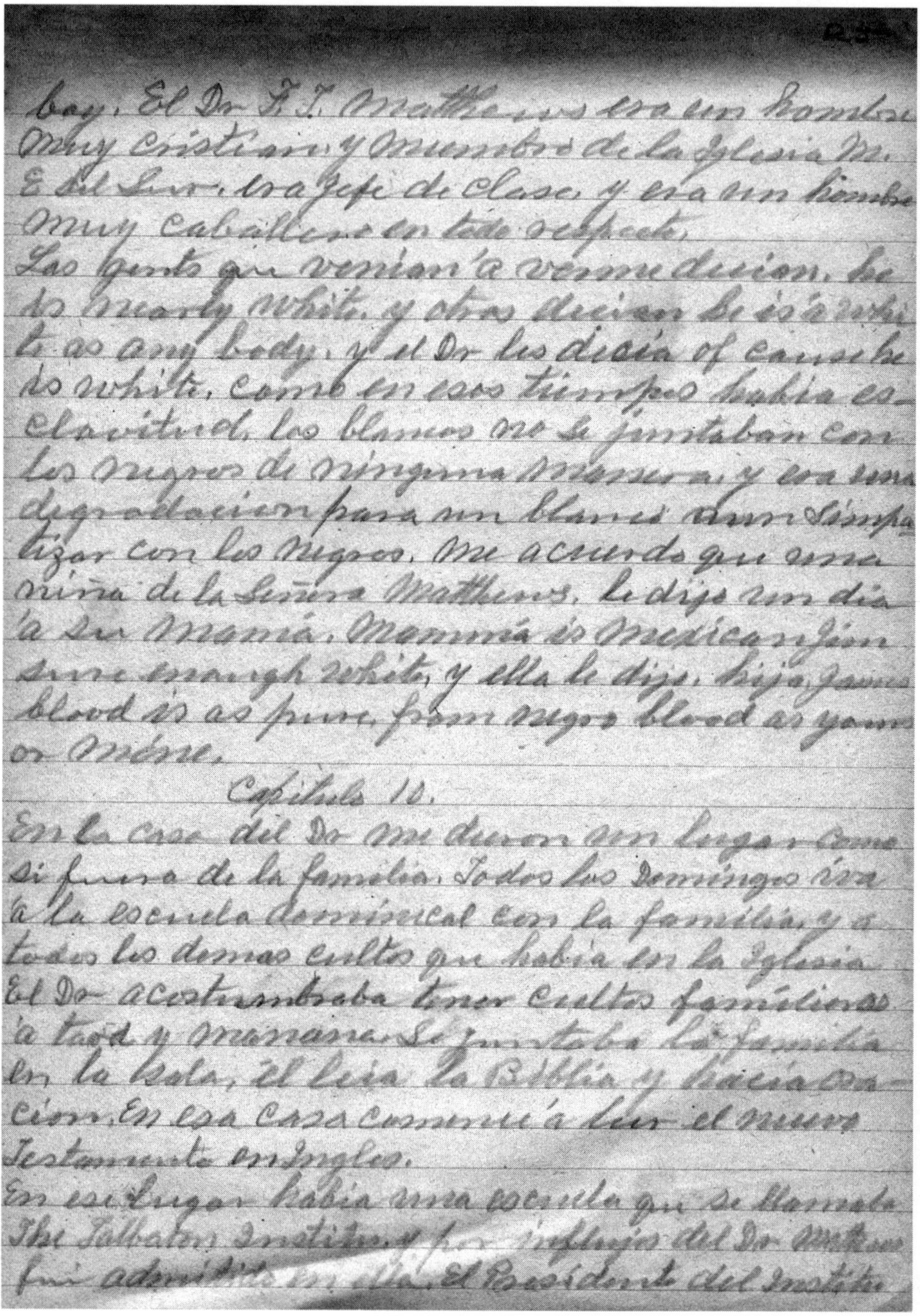

28

boy. El Dr. J. T. Matthews era un hombre
muy cristiano y miembro de la Iglesia M.
E. del Sur. era Jefe de Clase y era un hombre
muy caballero en todo respecto.
Las gentes que venian a verme decian. he
is nearly white. y otros decian he is a whi
te as any body. y el Dr les decia of course he
is white. como en esos tiempos habia es-
clavitud. los blancos no se juntaban con
los negros de ninguna manera, y era una
degradacion para un blanco el simpa
tizar con los negros. Me acuerdo que una
niña de la Señora Matthews, le dijo un dia
a su Mamá. Mamma is mexican Jim
pure enough white, y ella le dijo. hija, James
blood is as pure, from negro blood as yours
or mine.

Capitulo 10.

En la casa del Dr me dieron un lugar como
si fuera de la familia. Todos los Domingos iba
a la escuela dominical con la familia, y a
todos los demas cultos que habia en la Iglesia
El Dr acostumbraba tener cultos familiares
a tarde y mañana. Se juntaba la familia
en la sala, el leia la Biblia y hacia ora-
cion. En esa casa comencé a leer el nuevo
Testamento en ingles.
En ese lugar habia una escuela que se llamaba
The Salvator Institute. y por influjos del Dr Mathews
fui admitido en ella. El Presidente del Instituto

One of the original manuscript pages written by Santiago with pencil on primary tablet, in which he describes the people coming to see the "Mexican boy" and describes Mrs. Matthews' daughter's inquiry as to whether he was White.

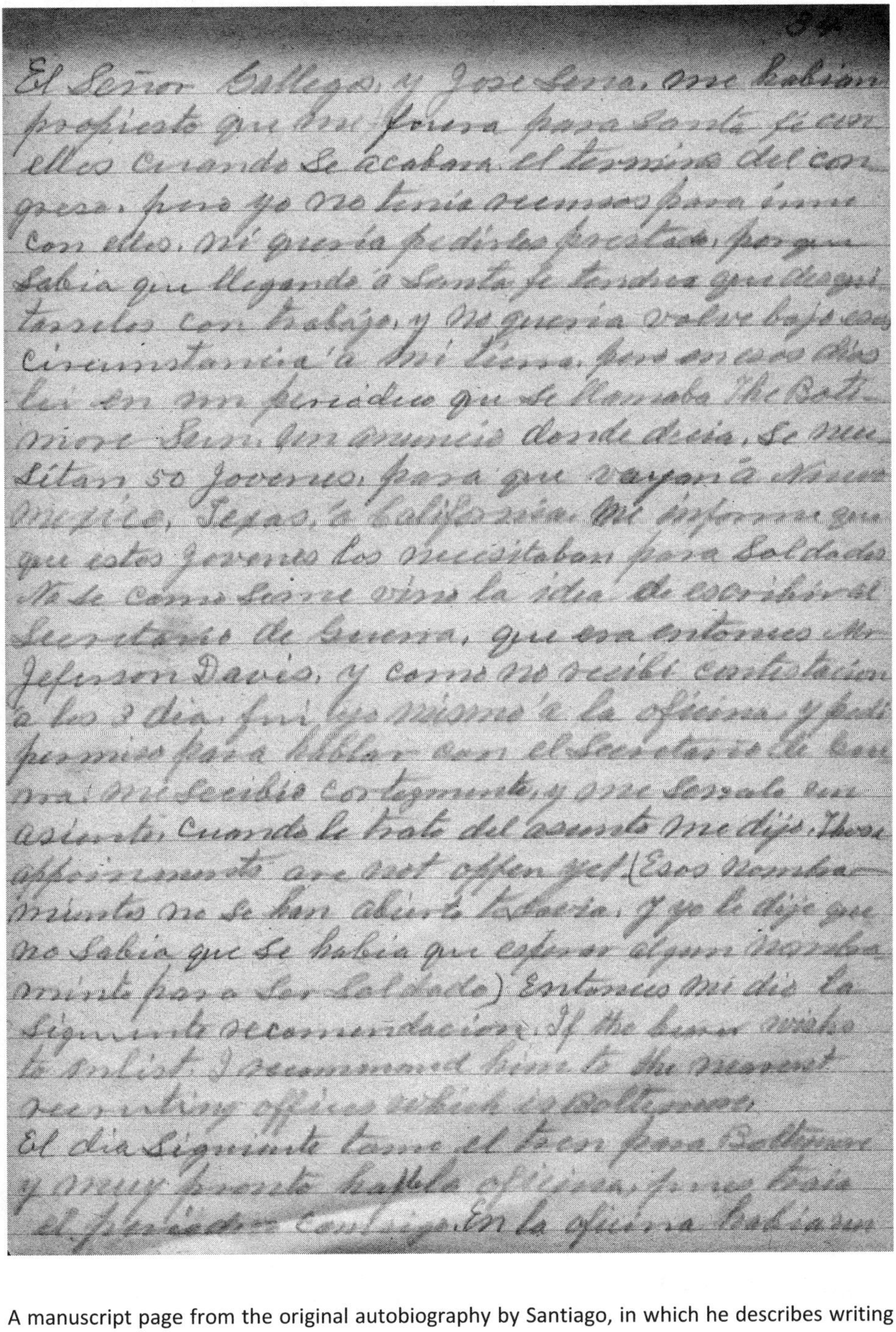

34

El Señor Gallego, y José Serna, me habían propuesto que me fuera para Santa fe con ellos cuando se acabara el termino del congreso, pero yo no tenía recursos para irme con ellos, ni quería pedirles prestado, porque sabía que llegando á Santa fe tendría que desquitarselos con trabajo, y no quería volver bajo esa circunstancia á mi tierra, pero en esos días leí en un periódico que se llamaba The Baltimore Sun, un anuncio donde decía, se necesitan 50 jovenes, para que vayan á Nuevo mejico, Texas, ó California. me informe que que estos jovenes los necesitaban para soldados No se como se me vino la idea de escribirle al Secretario de Guerra, que era entonces Mr. Jefferson Davis, y como no recibí contestación á los 3 días, fui yo mismo á la oficina y pedí permiso para hablar con el Secretario de Guerra. me recibió cortezmente, y me señaló un asiento, cuando le trate del asunto me dijo. Those appointments are not open yet (Esos nombramientos no se han abierto todavía, y yo le dije que no sabía que se había que esperar algun nombramiento para ser soldado) Entonces me dio la siguiente recomendacion. If the bearer wishes to enlist, I recommend him to the nearest recruiting offices which is Baltimore.

El día siguiente tomé el tren para Baltimore y muy pronto hallé la oficina, pues traía el periódico conmigo. En la oficina habían

A manuscript page from the original autobiography by Santiago, in which he describes writing to and meeting with U.S. Secretary of War, Jefferson Davis.

17, NUMBER 45 — ESTABLISHED AS THE CHRISTIAN ADVOCATE IN 1847 — Published Weekly · Subscriptions $3.00 Year — Second Class Posta

Conferencia Fronteriza Mexican sesionando en "La Trinidad" San Antonio, Texas.-1900 Presidio el Sr. Obispo W. A. Chandler

Conferencia Fronteriza Mexicana sesionando en "La Trinidad," San Antonio, Texas - 1900 - presidió el Sr. Obispo W. A. Chandler.

LA PRESENTE foto con que engalanamos este número hacemos reminiscencias de nuestro padres en el evangelio glorioso que aún predicamos; todos han sido llamados a descansar, pero sus obras les siguen; en la segunda fila y en el centro se destaca nuestro personaje el Rev. JOSE POLICARPO "POLLY" RODRIGUEZ, quién escribió su propia biagrafiá a petición del Dr. D. W. Carter quien fue superintendente del trabajo entre los mexico-americanos; "Polly" es el que aprece de pié y de barba blanca, y ahora que estamos por celebrar el primer centenario del trabajo metodista en Texas., nos sirve de marco, esta interesante foto. "Polly" según su propia biografía, se presenta como un Scout, Un soldado, incansable que peleo contra los indios, tomó parte en la guerra civil y por ultimo después de su conversión que se devió al mensaje de su primo Santiago Tafolla, que después, también Tafolla y "Polly" fueron ministros metodistas, que predicaron el evangelio en el estado de Texas. Debo esta información a una Revista antigua obsecuiada por el Prof. Fidel L. Tafolla.--A.M.G.

Conferencia Fronteriza Mexicana 1900 in San Antonio.

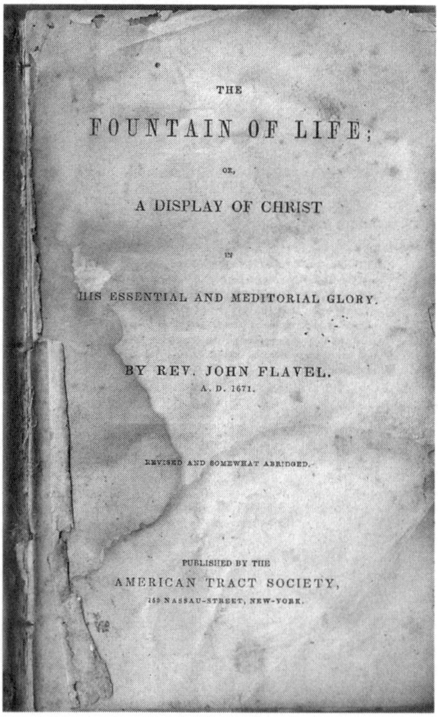

THE

FOUNTAIN OF LIFE;

OR,

A DISPLAY OF CHRIST

IN

HIS ESSENTIAL AND MEDITORIAL GLORY.

BY REV. JOHN FLAVEL.

A. D. 1671.

REVISED AND SOMEWHAT ABRIDGED.

PUBLISHED BY THE

AMERICAN TRACT SOCIETY,

150 NASSAU-STREET, NEW-YORK.

One of Santiago's books, *The Fountain of Life,* first published in 1671, by Rev. John Flavel, which was still in Santiago's possession when he died. (Courtesy of C. Tafolla.)

Anastacia and her sister, Sabinita, in their nineties.

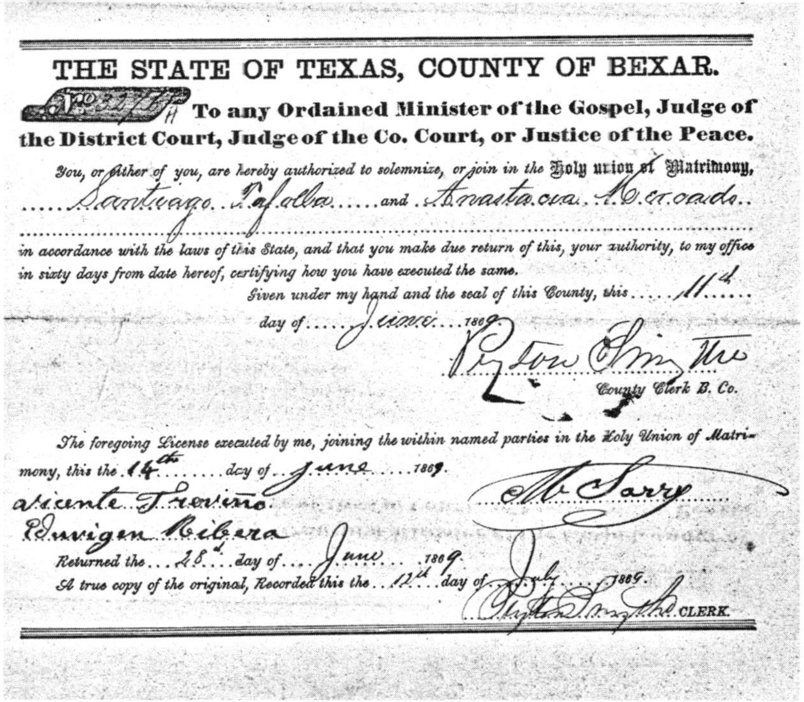

THE STATE OF TEXAS, COUNTY OF BEXAR.

No. 3171 H To any Ordained Minister of the Gospel, Judge of the District Court, Judge of the Co. Court, or Justice of the Peace.

You, or either of you, are hereby authorized to solemnize, or join in the Holy union of Matrimony, ...Santiago Tafolla.....and ..Anastacia Mercado..

in accordance with the laws of this State, and that you make due return of this, your authority, to my office in sixty days from date hereof, certifying how you have executed the same.

Given under my hand and the seal of this County, this.....11th.....

day of.....June.....1869.

Peyton Smythe

County Clerk B. Co.

The foregoing License executed by me, joining the within named parties in the Holy Union of Matrimony, this the 14th......day of...June....1869.

Vicente Treviño

Eduvigen Ribera

M. Sarry

Returned the....28th...day of....June.....1869.

A true copy of the original, Recorded this the..12th...day of...July.....1869.

Peyton Smythe CLERK.

Marriage Certificate of Santiago and Anastacia

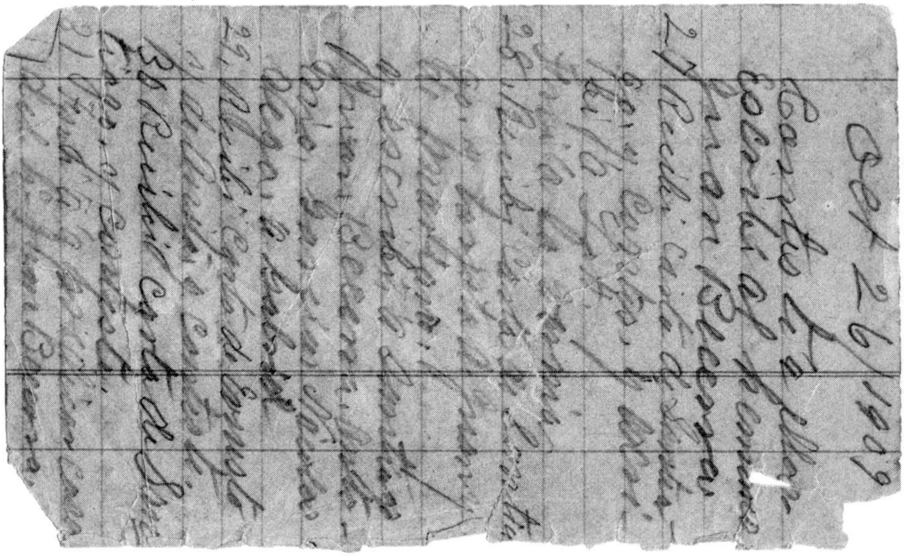

Oct 26/1909

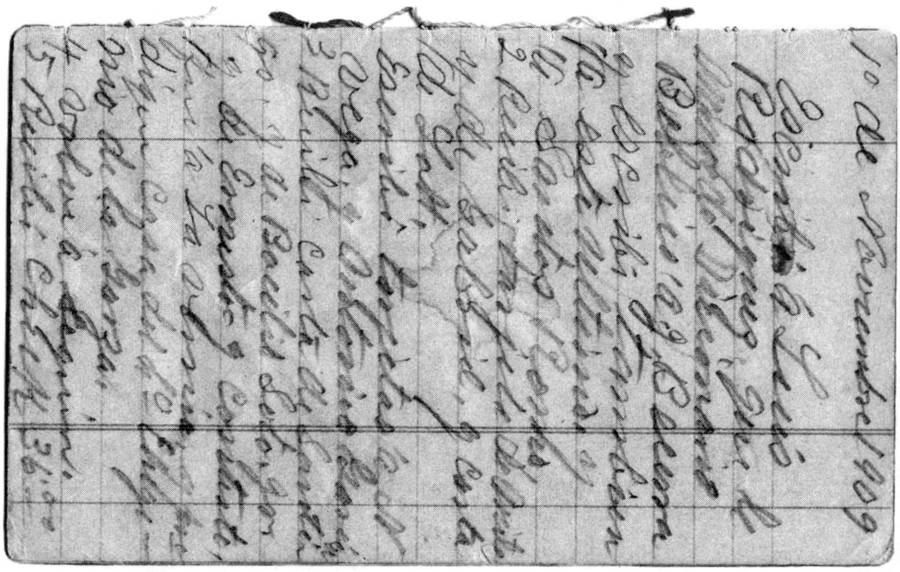

1° de Noviembre 1909

Pocket Journal of Santiago Tafolla, starting 10-1-1909.

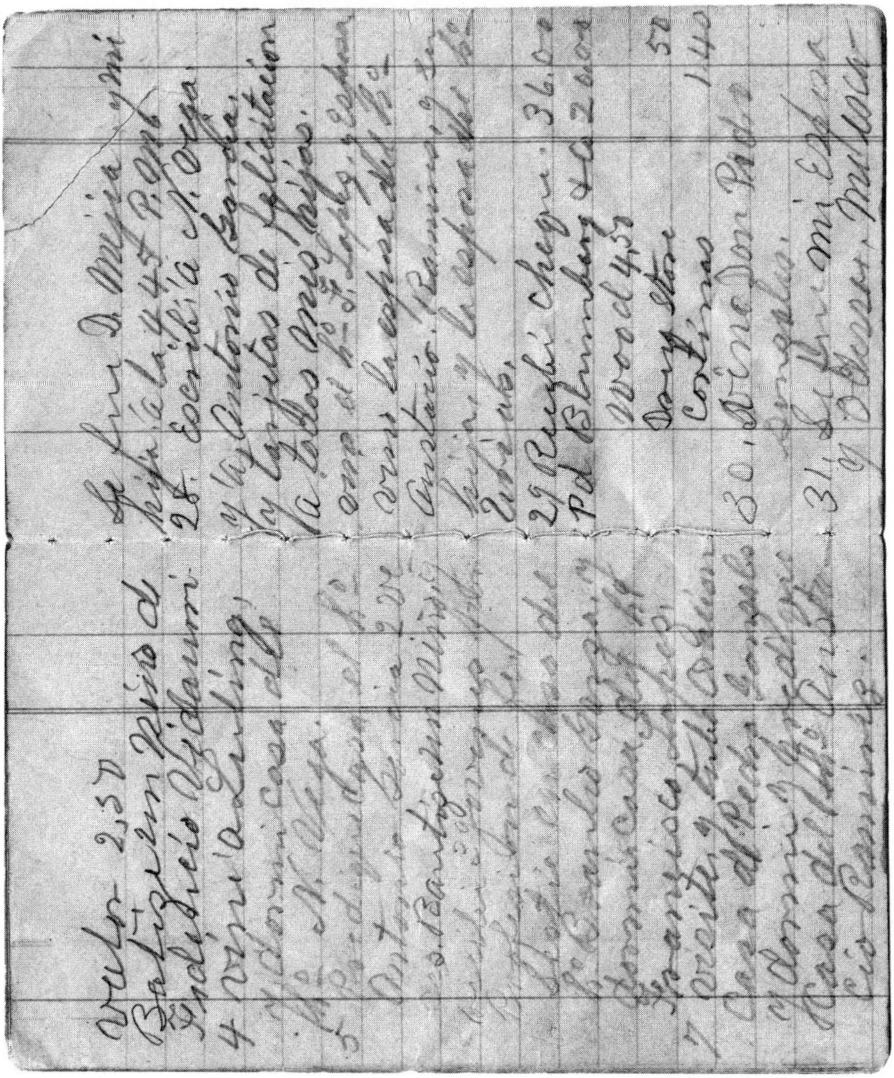

Pocket Journal of Santiago Tafolla, starting 10-1-1909.

O. W. and N. Div. BOUNTY LAND 3—1852. 13 Ex'r.

Invalid Orig. No. 338592
James Tafolla
Private – Capt. Palmus
2 US Cav
(1855-60)

THE ABOVE TITLE SHOULD BE INDORSED ON EVERY PAPER RELATING TO THIS CLAIM.

Department of the Interior,

BUREAU OF PENSIONS,

Washington, D. C., April 3, 1903

Sir:

Your claim No 338592 for Bounty Land under the Act of March 3rd 1855 has been [illegible] and will be examined at an early date and the result communicated to you.

In any communication you may find it necessary to address the Bureau relative to your claim please state the number of the same and the service upon which it is based.

Very respectfully
J. L. Davenport
Acting Commissioner

Mr James Tafolla
1345 Castro St
San Antonio
Texas.

Old War and Navy Division. 3—1883. O. L. G.

Sur. Orig. 7932
James Tafolla,
Indian War.

DEPARTMENT OF THE INTERIOR,

BUREAU OF PENSIONS,

WASHINGTON, D. C., Oct. 10, 1904.

Sir:-

You are advised that your above cited claim for pension under the acts of July 27, 1892 and June 27, 1902 is rejected on the ground that you did not serve thirty days in the Texas and New Mexico Indian War prior to January 1, 1856, or in any other war mentioned in the acts under which pension is claimed.

Your claim for bounty land is rejected on the ground of no title for the reason that your entire service was rendered subsequent to March 3, 1855.

Very respectfully,
E. F. Ware
Commissioner.

Mr. James Tafolla,
831 West Ruiz Street,
San Antonio, Texas.

Correspondence with the War Department and Dept of Interior, regarding pension: 10-10-04 to 1-14-11.

Old War and Navy Division

3—1883.

No. 338,592
James Tafolla.
Bounty Land.

DEPARTMENT OF THE INTERIOR,
BUREAU OF PENSIONS,
WASHINGTON, D. C., April 3, 1906.

Sir:-

In reply to your communication of the 14th ultimo, received on the 19th, relative to your above cited claim for bounty land, you are again advised that said claim was rejected October 10, 1904, on the ground of no title for the reason that your entire service was rendered subsequent to March 3, 1855.

Service to give title to bounty land must have been rendered prior to March 3, 1855, and there is no law under which a bounty land warrant could be issued for service rendered subsequent to that date.

Very respectfully,
V. Warner.
Commissioner.

Mr. James Tafolla,
San Antonio, Texas.

Two inclosures

1730652

WAR DEPARTMENT,
THE ADJUTANT GENERAL'S OFFICE,
WASHINGTON,

January 4, 1911.

Mr. James Tafolla,
Post Office Box 245,
Seguin, Guadalupe County, Texas.

Sir:

Referring to your application for a certificate in lieu of destroyed discharge certificate as a member of Company D, 2d United States Cavalry, I am directed by the Secretary of War to inclose herewith the desired certificate, and to return the affidavit of Gustav Frasch, which appears to relate to a claim for pension in this case.

Very respectfully,
The Adjutant General.

Correspondence with the War Department and Dept of Interior, regarding pension: 10-10-04 to 1-14-11.

354 THE ROYAL GALLERY.

SMALL BEGINNINGS.

A TRAVELER through a dusty road strewed acorns on the lea,
And one took root and sprouted up, and grew into a tree.
Love sought its shade, at evening time, to breathe its early vows:
And age was pleased, in heats of noon, to bask beneath its boughs;

He walled it in, and hung with care a ladle at the brink,
He thought not of the deed he did, but judged that toil might drink.
He passed again, and lo! the well, by summers never dried,
Had cooled ten thousand parching tongues, and saved a life beside.

"It stood a glory in its place,
A blessing evermore."

The dormouse loved its dangling twigs, the birds sweet music bore;
It stood a glory in its place, a blessing evermore.

A little spring had lost its way amid the grass and fern,
A passing stranger scooped a well, where weary men might turn;

A dreamer dropped a random thought; 't was old, and yet 't was new;
A simple fancy of the brain, but strong in being true.
It shone upon a genial mind, and lo! its light became
A lamp of life, a beacon ray, a monitory flame.

A page from *The Royal Gallery,* a poetry anthology belonging to Santiago. The front cover is missing, but this is possibly *The Royal Gallery of Hampton Court Illustrated* published in 1898. (Courtesy of C. Tafolla.)

MAIL DIVISION. 3—977.

, Clerk.

Department of the Interior,

BUREAU OF PENSIONS.

MAR 16. 1904

SIR:

Receipt of the paper.... below designated is acknowledged, as of the date stamped above

1. Affidavit Gustav Trasch
2.
3.
4.
5. James Tafolla
6. Ind. war

Very respectfully,

E. F. Ware

Commissioner.

N. B.—Give name and service of soldier and number of claim when writing to this Bureau.

10707b200m11-03 0-16

B

Old War and Navy Division. 3—528. (Old No. 3—609.)

DEPARTMENT OF THE INTERIOR,

BUREAU OF PENSIONS,

WASHINGTON, D. C., March 10, 1903

Sir

Your claim for pension under the ~~Act July 27, 1892~~ Act June 27, 1902, has been received and given the following title:

INDIAN WARS.

SURVIVOR Orig. No. 7932

James Tafolla alias Santiago Tafolla

The claim will be considered when reached in its order, and the requirements, if any, duly communicated to you. The title given above should be indorsed on every paper relating thereto which may hereafter be filed in this Bureau.

Very respectfully,

B. F. Ware

0-4 Commissioner.

OCT 29 1904

OFFICE OF
MILO B. STEVENS & CO., *Washington, D. C.,* 190
Solicitors of Claims and Patents.

We wrote you for additional information in ~~your~~ claim for Bounty Land. Your answer does not appear to have been received. We will take prompt action as soon as we hear from you. Jas Taffallo—alias Santigo

Milo B. Stevens & Co.

Attorneys, Etc.

Correspondence with the War Department and Dept of Interior, regarding pension: 10-10-04 to 1-14-11.

3-1-1910

HE SAN ANTONIO DAILY

She Lived for Most Part in San Antonio

MRS. ANITA TAFOLLA MEJIA.

Mrs. Anita Tafolla Mejia, 22 years old, wife of Dolores Mejia, 1012 West Houston Street, died in her home at 3:20 o'clock Monday afternoon after an illness of eight days. She was born in San Marcos, but had made San Antonio a place of residence a greater portion of her life. Her first schooling was in the Laredo Seminary, but she graduated in 1905 from the San Antonio Female College. She was for several months assistant teacher in the McDonald Institute, Durango, Mexico. There survive, besides the husband and 1-month-old baby, the father, mother and several brothers and one sister.

The father, Rev. James Tafolla, is one of the pioneer residents of the State, having engaged in the early days in many fights with the Indians. He has been actively engaged in missionary service in Texas and Mexico for the past thirty-three years, and now holds a pastorate in Seguin.

The brothers are J. J. Mercado, professor of Spanish in the Houston high school; Pete Tafolla, San Antonio; Gabriel Tafolla, Friotown; James Tafolla Jr., San Antonio; Marion Tafolla of San Antonio, Lott Leo Tafolla of Houston and Rev. Ernest A. Tafolla of Austin. The sister is Mrs. Mollie Tafolla Mendez of San Antonio.

The funeral services will be held from the Mexican Methodist Church, intersection of South Pecos and San Fernando Streets, Tuesday afternoon at 4 o'clock. Interment will be made in City Cemetery No. 3.

PRESS: TUESDAY MORNI

FUNERAL OF MISSIONARY

Last Sad Rites Over Body of Rev. E. A. Tafolla Observed Yesterday.

REV. E. A. TAFOLLA.

The funeral of Rev. E. A. Tafolla, the Mexican Methodist missionary, who was a native of San Antonio, took place yesterday afternoon.

Obituaries for Anita Tafolla Mejía, and Rev. Ernesto Tafolla, both 1911, one mismarked 3-1-1910, by hand.

REV. SANTIAGO TAFOLLA

Oja de Servicio

Según la biografía del Rev. Policarpo Rodríguez el hermano Tafolla era su primo, era oriundo de New México y se había educado en Georgia. Fue soldado donde sirvió como clarín. Después de la Guerra Civil se estableció en Bandera o cerca de la comunidad de Prvilegio donde vivía el hermano Rldríguez.

En ese tiempo, después de la Guerra Civil, se convirtió, talvez en el año 1875 y según parece era miembro de la Misión de Bandera de donde fue recomendado a la Conferencia Anual que se reunió en 1876, en Segúín en Octubre 25 de ese año, para ser admitido a Prueba. Esta Conferencia era la West Texas que ahora es la Southwest Texas Conference.

En esa Conferencia fue mandado a Laredo como su primer cargo pastoral.

OJA DE SERVICIO

1876-1877---Laredo, Texas

1877-1878---Corpus Christi, Texas

1878-1879---Corpus Christi, Texas

1879-1880---Corpus Christi, Texas

1880-1881---San Diego Mexican Mission District-Presbítero Presidente

1881-1882--- " " " " " " "

1882-1883 " " " " " " "

1883-1884---Guadalupe Mexican Mission

1884-1885---San Antonio ~~Mexican~~ District-Presbítero Presidente

1885-1886--- " " " " " "

1886-1887--- " " " " " "

1887-1888--- " " " " " "

1888-1889----" " " " " "

(1885–1889: Conferencia Fronteriza)

1889-1890----San Antonio, La Trinidad

1890-1891----Monterrey ~~Mexican~~ District, Presbítero Presidente

Santiago's "Hoja de Servicio."

EL EVANGELISTA MEXICANO

EL REV. SANTIAGO TAFOLLA

En unión al numeroso círculo de sus conocidos y amigos, yo también he experimentado un profundo dolor en mi corazón por la noticia del fallecimiento de este buen hermano e inolvidable amigo; y no puedo menos que decir algo de él en las columnas de El EVANGELISTA MEXICANO.

Durante nuestra larga asociación en los trabajos evangélicos, el me platicaba mucho de su historia.

Fue oriundo de Santa Fe, capital de Nuevo México. Si es como dijo Jeremías "bueno es que uno llevara el yugo desde su mocedad", entonces era bueno con el sujeto de esta historia, porque él me ha dicho que desde su niñez, sus padres, o los que tenían el encargo de él, solían mandarlo a las montañas para cortar leña a la madrugada, aun caundo la nieve cubría la tierra.

En este tiempo un oficial del ejército americano, viendo que era un muchacho vivo, industrioso y bueno, quiso llevárselo a casa, en el lejano Estado de Georgia, Estados Unidos; y en efecto, lo hizo.

Allá tuvo la oportunidad de educarse en las escuelas del país, y de aprender inglés y otras cosas útiles, aun el oficio de zapatero.

Asistía a los cultos de la iglesia Metodista y fue verdaderamente convertido a Dios.

Más tarde cambió de residencia a la gran ciudad de Baltimore, cerca de la ciudad de Washington, capital de la nación.

Ya de una edad suficiente para alistarse en el ejército, se le vino la idea de hacerse soldado, probablemente con la esperanza de volverse a su tierra natal. Así es que se fue al despacho en la ciudad de Baltimore donde hacian reclutas para el ejército.

El militar encargado no quiso admitir su solicitud por ser él de tan pequeña estatura, y con una sonrisa le despidió.

Esto ofendió en gran manera la suceptibilidad de Santiago, y con la independencia y energía que le eran características siempre, se puso en camino para Washington, y allí con toda franqueza se metió en el despacho mismo del Ministro de Guerra quien en aquel tiempo era el Sr. Jefferson Davis, y que más tarde fue Presidente de los Estados Confederados del Sur.

Este distinguido caballero escuchó con su acostumbrada finura y bondad a nuestro joven Tafolla, quien francamente le contó lo que le había pasado con el militar en Baltimore; y le preguntó, que siendo tan jóven y pequeño que podría hacer? "Señor," le dice Tafolla, "yo puedo andar a caballo, puedo pelear y puedo tocar el clarín."

Muy divertido el Ministro de Guerra con la respuesta del joven, le dio una nota que llevara al oficial de Baltimore, en que ordenaba que Santiago Tafolla fuese reclutado como soldado en el ejército americano.

Poco tiempo después fue enviado con otros a las lejanas fronteras de Texas, y fue hecho el primer clarín del regimiento. Este fue uno de los más famosos regimientos de todo el ejército. Su coronel era el Sr. Albert Sidney Johnston, más tarde el famoso general del ejército del Sur, que perdió su vida en la batalla de Shiloh. Su teniente coronel vino a ser el todavía más renombrado, Robert E. Lee, generalísimo de los ejércitos del Sur.

Muchos otros oficiales de este regimiento llegaron a ser muy distinguidos generales ora en el ejército del Norte, ora en el del Sur.

Otras muchas cosas interesantes de su vida militar me contó el hermano Tafolla, pero creo que no son del caso referir aquí.

Salido del ejército se estableció en el condado de Bandera, Texas, casándose con una prima de su buen amigo D. José Policarpo Rodríguez, conocido de todos como Polly Rodríguez.

Este, más anciano que Tafolla, entro al ministerio un año más tarde, y sigue hasta la fecha tan consagrado como siempre, y en la actualidad desempeña el oficio de Colportor de la Sociedad Bíblica Americana, entre los mexicanos de Texas.

Hubo hijos e hijas de fruto de este primer enlace matrimonial. Muerta esta digna senora, más tarde el hermano Tafolla se casó con la señora Anastacia Vda. de Mercado, madre de nuestro querido hermano Juan J. Mercado, mucho tiempo ministro y miembro de la Conferencia Fronteriza Mexicana, y en la actualidad profesor en el colegio de Houston, Texas. De este matrimonio nacieron varios hijos y una hija.

Algunos de ellos precedieron al padre a la eternidad, y los demás permanecen en este valle de lágrimas, acompañando a la viuda madre en su acerbo dolor; ella por la pérdida del mejor de los maridos, y ellos por la del más tierno y amante de los padres.

Durante su vida en el ejército, perdió su experiencia cristiana y se volvió a vivir la vida del mundo. Pero no se olvidó Dios de él. En Diciembre de 1874, la Conferencia Occidental de Texas, (West Texas Conference) se reunió en González, Texas y fue organizada la obra misionera entre los mexicanos de las fronteras de Texas y México, y él que esto escribe, fue nombrado su Presbítero Presidente, con tres predicadores mexicanos.

Estos tres predicadores recibieron su licencia de predicar y su recomendación para admisión a prueba en la conferencia Anual de la Conferencia Trimestral de la Congregación Americana de Corpus Christi, Texas, de la cual yo era Pastor.

La siguiente Conferencia Anaual, la de 1875, se verificó en San Antonio, Texas siendo considerable el aumento en el número de predicadores para admisión a prueba.

"El Rev. Santiago Tafolla", in *El evangelista mexicano* by A.H. Sutherland.

Entre éstos estuvo el hermano Jose Maria Casanova, Este fue nombrado para la nueva mision que quise formar en San Antonio, Texas, donde me quedé más de un mes después de la Conferencia Anual para dar principio a estos trabajos. El éxito de la misión en San Antonio, fue muy grande desde un principio.

Se hizo notorio este movimiento entre los mexicanos por todas partes, Muchos vinieron de los alrededores para ver que se hacía.

Los sacerdotes romanos se hicieron solidarios y hasta ruidosos en su contra que sólo servía para llamar más la atención y así extender la buena obra.

De entre los nuevos convertidos, Dios levantó varios jovenes bastante inteligentes y muy llenos de fervor para la causa, siendo el más hábil y activo, uno que se llama Trinidad Armendariz.

Yo le llamaba en aquel tiempo mi Lutero. No había quien resistiera la fuerza y valor de su palabra. Aun tenía discusiones públicas con sacerdotes y otros campeones de la iglesia Católica, pero pronto estos supieron dejarle en paz; ni uno de ellos tuvo el atrevimiento de discutir con él por segunda vez.

En este tiempo supe por fuente fidedigna de un "joven" como le decían en Bandera, que se interesaba en la novedad, que entonces llamba tanto la atención de los mexicanos, y escribí al hermano Jose María Casanova que fuera a hacerle una visita.

Así lo hizo, y no sólo una vez; también extendiendo Armendáriz sus trabajos hasta allí. Este "joven" era el hermano Santiago Tafolla. Pronto recobró su perdida experiencia cristiana y sirvió de base para la obra entre los mexicanos de su vecindario que se llamaba Privilegio, y del cual el señor D. Policarpo Rodríguez era el gran cacique.

Tafolla se unió con la iglesia Americana de Bandera, y se sintió llamado por Dios a predicar el evangelio. Su Conferencia Trimestral le dio su licencia y su recomendación a la Conferencia Anual.

Vino a ésta que se reunió en Octubre de 1876 en Seguín, Texas. Allí con gran gozo se juntó con los predicadores mexicanos que yo tenía. En sus pláticas con ellos supieron que él no se había bautizado después de hacerse protestante, y le expusieron sus razones por qué el bautismo católico romano no se debe reconocer como bautismo cristiano, y vino él para consultar conmigo sobre este asunto y yo acabé de convercerle que ellos tenían razón.

El entonces me pidió el bautismo, el cual se efectuó ante la Conferencia Anual, el Obispo Doggett, Presidiendo, luego él, Gumersindo Paz; Trinidad Armendáriz, y otros varios fueron recibidos a prueba en la Conferencia Anual.

Esto fue en Octubre de 1876 en Seguín, Texas, efectivamente 35 años antes de su muerte en el mismo lugar y mes.

Su primer nombramiento fue para Laredo, Texas, a donde llegó en Marzo de 1877. Yo había estado en Laredo, Texas y en Nuevo Laredo, Tamaulipas, en Septiembre de 1875 y tenido varias reuniones, de modo que estaba algo preparado el terreno para este nuevo y fiel obrero.

A pocas semanas de su llegada yo le hice mi primera visita, encontrando que ya se habían empezado a reunir con él y con su familia algunos mexicanos.

Haciamos invitaciones a todo el pueblo y tuvimos buenas reuniones durante aquellos días, y de allí en adelante, no han faltado bastantes evangélicos en Laredo.

A fines del año 1877, la Conferencia se convocó en Corpus Christi, Texas y el hermano Tafolla fue cambiado de Laredo a otra parte, y otro predicador enviado a Laredo.

Este fue el principio de una vida ministerial que duró exactamente 35 años acabándose en el mismo mes y en el mismo lugar donde comenzó; una vida ministerial de singular consagración y pureza.

Su conocimiento del inglés le fue muy ventajoso, porque así es que pudo asociarse y conversar con algunos de los antiguos predicadores americanos que en aquellos tiempos pertenecían a la Conferencia Anual y que querían mucho al hermano Tafolla y al Superintendente de las misiones mexicanas y que se complacían en ayudarme hasta donde podían en la ardua obra que yo llevaba a cabo.

Muchos y largos viajes y duros trabajos hemos pasado juntos el hermano Tafolla y yo, por toda esa frontera del Norte, tanto en Texas como en México. Dios bendecía la obra de nuestras manos y nosotros nos recocijamos en una obra tan bendecida. Este hermano fue nombrado Presbítero Presidente de un distrito en la Conferencia de 1881 en Goliad, Texas; siendo, creo, el primer mexicano en la historia de la Iglesia que ocupara tal puesto.

Muchas veces después lo ocupaba, y siempre con honra, éxito y buena aceptación. El recuerdo de aquellos tiempos y de aquel querido hermano, será siempre, para mi corazón, muy grato. Que la obra de él para sus paisanos, una obra tan espiritual y limpia, sea conservada y perpetuada para siempre, es de desearse. El ha sido un modelo digno de estudio y de imitación de parte de los jóvenes predicadores mexicanos.

En cuanto a los bienes de este mundo, era pobre y no pudo dejar riquezas materiales a sus hijos, pero les ha dejado la incalculable riqueza de una historia sin mancha, de un carácter noble y honrado. Les ha dejado una estela de luz que señala el derrotero del cielo. Esta es una herencia que bien pocos han podido dejar a sus hijos.

Esto es muy precioso y ellos no debieran menospreciarlo, sino enterarse de todo su valor. ¡Oh, que de entre ellos se levanten otros todavía movidos por el Espíritu Santo, para tomar los lugares vacantes por la muerte reciente de su buen padre y su querido hermano Ernesto.

Así mismo pido a Dios que consuele y santifique cabalmente por medio de su Santo Espíritu a nuestra querida hermana Anastacia, que llora la separación de su compañero de su vida, a fin, que al cruzar ella también el río, encuentre del otro lado a su querido compañero en esta vida, esperándola en unión de los muchos de su familia que ya lo han pasado.

Y a vosotros, oh hermanos veteranos de las dos Conferencias Anuales, la Fronteriza y la Noroeste, habéis visto otra vez desaparecer de entre vosotros uno de los más antiguos, útiles y honrados compañeros nuestros, que es un aviso a vosotros para que estéis apercibidos, y listos, por que habéis visto que cuando menos se espera quizás, oiréis la llamada.

Y habéis visto que una muerte como la de él, es recompensa suficiente por todo lo que podrá sufrir uno por la bendita causa del maestro.

Sed fieles hasta la muerte y recibiréis la Corona inmarcesible de la vida eterna.

Vuestro amigo y compañero.

Alejandro H. Sutherland.

EL EVANGELISTA MEXICANO

MUERTE DE UN VETERANO EN LA FRONTERA

En el cumplimiento del deber y exhortando a los miembros de la Conferencia Trimentral, reunida en Seguín, Texas, el domingo 29 del mes próximo pasado, falleció el Reverendo Santiago Tafolla.

No fue víctima de ninguna enfermedad por la que exhalara ni una queja. Muy ajeno a las presiones morales que causa el contacto con los deberes familiares, y cuando más empeñado estaba en infundir calor espiritual a los miembros de su iglesia; lleno de vida y elevada su alma a las regiones de lo desconocido, cayó el valiente predicador del Evangelio de quien puede decirse que no vio o no sintió ni cuando se le acercó la muerte. Cerró los ojos y cayó en los brazos de su Presbítero, nuestro muy apreciable hermano el Dr. Carter, y de allí a San Antonio, donde se le hicieron los funerales.

El hermano Tafolla elevó sus preces al Eterno , pidiéndole que le concediera morir en el cumplimiento de sus deberes o precisamente cuando estuviera predicando.

Dios escuchó la súplica de su siervo y lo llevó de este mundo cuando estaba en el calor de la lucha. Cayó como caen los valientes, admitiendo como el apóstol que el morir es ganancia.

Había sido un valiente antes de su conversión, cuando en compañía del Rev. J. P. Rodríguez y con las armas del gobierno Americano se lanzó a luchar con los indios que por aquella época infestaban la frontera de Texas.

Tanto Polly Rodríguez como Tafolla, sacaron de aquella lucha, que algunas veces fue cuerpo a cuerpo, la manera de probar su valor. Una vez convertido el hermano Tafolla, no perdió esa virtud, sino que la cultivó haciendo proezas en honor de su Maestro y Señor.

Por el año de 1876, comenzó a predicar en Bandera y el mismo año, se conectó a la obra misionera. Laredo fue el primer lugar que se le asignó como itinerante y partió con todo y familia en un mueble de atraccion animal, pues el ferrocarril no llegaba aun a la ciudad ribereña.

En aquella época la Iglesia Metodista Espiscopal del Sur, en su trabajo entre mexicanos, estaba auspiciada por la Conferencia Occidental de Texas. En San Antonio se reunió la Conferencia Anual el año de 1875 en virtud del crecimiento de la obra.

Digno de nuestra especial atención es la que aquel hombre de Dios, hizo antes de partir para Laredo a recibir su primer campo de trabajo evangélico.

La Conferencia no había hecho apropiación ninguna para gastos de viaje de sus predicadores y el hermano Tafolla, escaso de recursos para acometer tan ardua empresa, hipotecó sus propiedades en el condado de Bandera para hacerse de los recursos necesarios para su camino.

En su nueva conceción llenó algunos de los más importantes cargos pastorales habiendo servido tambien como presbítero presidente en varios distritos. Dos veces fue nombrado por su Conferencia Delegado clerical a la Conferencia General.

Para el Rev. Tafolla no había esos estorbos que se llaman inclemencias del tiempo, ni diques insuperables si se trataba de cumplir alguna promesa que hubiera hecho de llegar tal día y hasta a tal hora a predicar el Evangelio.

El que esto escribe, tuvo la oportunidad de saber por experiencia algo de lo que ahora escribe. En cierta ocasión, salimos de San Diego, para ir, según me dijo a cumplirles su palabra a unos hermanos.

La lluvia se desencadenó a extremo de que los arroyos no daban paso aparte de lo fatigoso del camino, cuando el agua caía a torrentes.

Al fin llegamos a un arroyo muy peligroso, por lo turbulento y lo profundo de su cauce. Se detuvo un poco, pensó algo, meditó y quiza se puso en oración, mientras yo contemplaba absorto aquel mundo de agua que parecía amenazar con la muerte al que se acercaba a contemplarlo.

De súbito el hermano Tafolla me dijo; hermano Grado, sabe usted nadar? Le contesté que sí. Pues entonces yo no puedo quedar mal. Yo le prometí al hermano X, que esta noche predicaría en su casa y sólo muerto no lo hago.

Diciendo y haciendo, tomó las riendas, endilgó a la bestia hacia la orilla y en ese momento sentí algo frío en todo el cuerpo, porque comprendí lo arriesgado de la empresa.

La bestia al entrar y no hallando plan, comenzó una lucha tremenda, no tan sólo por llegar al otro lado, sino por deshacerse del peso, lo cual consiguió.

La bestia, ya libre, salió sin dificultad, y el hermano Tafolla y yo quedamos en medio del peligro; peligro digo para mí, pues el se arrojó y con destreza admirable, no obstante el peso de la ropa y de las botas, salió ileso a tierra.

Considéreseme por un momento, solo, medio a medio de aquel monstruo, con sus amenazantes olas y animado únicamente con los gritos del veterano que me decía: tírese, no tenga miedo.

Para no hacer larga mi narración, les diré a los lectores que pasé, no digo que sin miedo, pero sí sin novedad. Pagó el hermano Tafolla a dos rancheros muy expertos en natación y éstos sacaron el mueble.

Nadie se puede imaginar la satisfacción que sentía aquel luchador, cuando sanos y salvos reanudamos nuestro camino para ir a cumplir lo prometido.

En otra ocasión, el hermano Tafolla, al despedirse de mi familia que radicaba en Mier, Tamaulipas, dijo a mi mamá: por aquí nos veremos dentro de tres meses, con el favor de Dios, el jueves de tal semana a las tres de la tarde.

Cumplido el tiempo, y el jueves que él citó, como un cuarto de hora antes de las tres, nos acordamos del hermano Tafolla, y no acababamos todavía de tratar de la promesa, porque sabíamos ya lo cumplido que era, cuando se paró en la puerta y nos dijo: Buenas tardes, hermanos.

Article "Muerte de Un Veterano" by Pedro Grado, in *El Evangelista Mexicano*, and accompanying poem "A la sentida muerte de Santiago Tafolla."

Muchas de estas cosas que nos indican al verdadero hijo de Dios, puedo relatar, pero por ahora sólo seguiré hablando de sus funerales.

A las diez de la manana los Revs. J. P. Rodríguez, C. A. Rodríguez, Pedro Grado, A. R. Cárdenas, Manuel Sanmiguel y Tarantino, nos encargamos de sacar el féretro de la casa numerada 1012 de W. Houston St. residencia de Santiago Tafolla, hijo del extinto.

Nos acompañaban como honorarios a nuestro oficio, los Revs. S.H.C. Burgin, B. A. Godbex, T. E. Harrison J. B. Young, C. B. Cross y Z. V. Liles, todos de la iglesia metodista. Abudaron Abundaron las coronas y las flores naturales.

En la capilla tomaron la palabra los hermanos J. P. Rodríguez, D. W. Carter, A. R. Cárdenas, Basilio Soto y E. B. Vargas.

Una vez en el panteón el Rev. Basilio Soto, reanudó la parte del ritual con lo que se terminó el sepelio del decano de la Conferencia Fronteriza Mexicana y uno de los obreros más consagrados y más celosos por la Causa de Nuestro Señor Jesu-Cristo.

¡Duerma en paz el campeón del evangelio! el intachable obrero y hermano sincero y cariñoso. Tus hermanos al recordar de ti, procuran seguir las huellas que con tu ejemplo les trazaste.

Pedro Grado

Hijos del Rev. Santiago Tafolla: Pedro Tafolla, James Tafolla, Ernesto Tafolla,

Mariano Tafolla, Lott Tafolla, Gabriel Tafolla

A LA SENTIDA MUERTE DEL REV.

SANTIAGO TAFOLLA

Cuando un valiente cae en la contienda
Por una noble causa,
Sonrie de placer, por que es un triunfo
Morir con esperanza.

Cuando un vidente de inspirado verbo,
Al golpe de la Parca
En este mundo cae, alla en el otro
Glorioso se levanta.

Cuando el sol ya se pierde en el ocaso
Del justo a la mirada,
Hay luto en nuestras almas y el contempla
Una nueva manana.

Cuando un heroe sucumbe y en su puesto
Yace junto a su espada,
Solo cambia de ambiente y su partida
Reputa por ganancia.

Por eso el irredento se prosterna,
Aun se humilla y calla,
Ante el cadaver del que a voz en cuello
La verdad proclamara.

Los que mueren creyendo la doctrina
De Aquel que al mundo salva,
Su carga depositan a la puerta
De su hogar y descansan.

No los ves, como ostentan satisfechos
Sus vestiduras blancas?
Lavadas fueron en la sangre regia
Del Cordero sin mancha.

Asi es el heroe que vencio en la lucha
Y mi canto proclama:
En sus cienes ostenta la corona
Y en su mano la palma.

Pedro Grado.

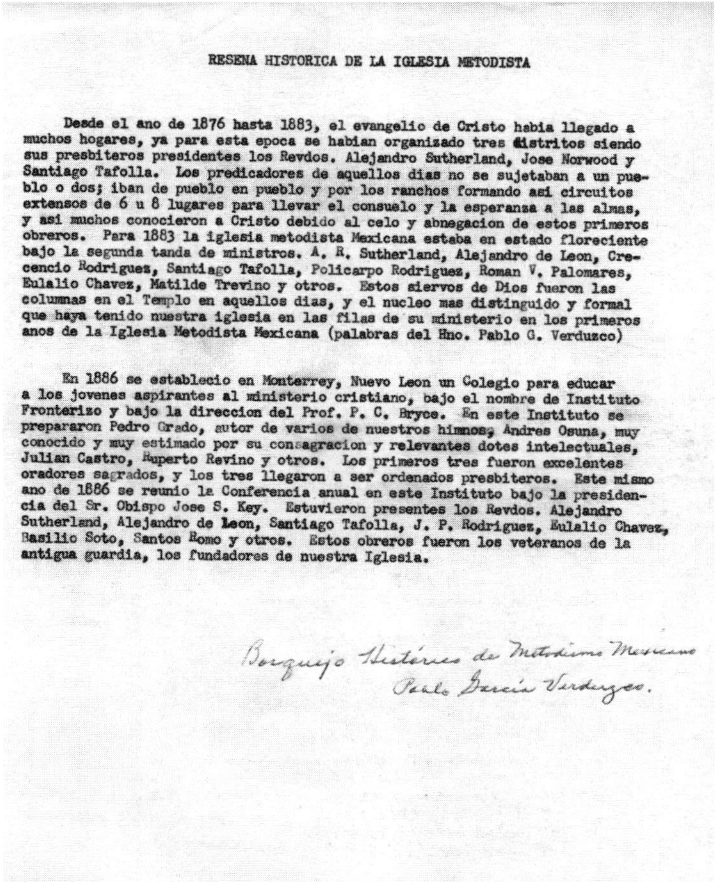

RESENA HISTORICA DE LA IGLESIA METODISTA

Desde el ano de 1876 hasta 1883, el evangelio de Cristo habia llegado a muchos hogares, ya para esta epoca se habian organizado tres distritos siendo sus presbiteros presidentes los Revdos. Alejandro Sutherland, Jose Norwood y Santiago Tafolla. Los predicadores de aquellos dias no se sujetaban a un pueblo o dos; iban de pueblo en pueblo y por los ranchos formando asi circuitos extensos de 6 u 8 lugares para llevar el consuelo y la esperansa a las almas, y asi muchos conocieron a Cristo debido al celo y abnegacion de estos primeros obreros. Para 1883 la iglesia metodista Mexicana estaba en estado floreciente bajo la segunda tanda de ministros. A. R. Sutherland, Alejandro de Leon, Crecencio Rodrigues, Santiago Tafolla, Policarpo Rodriguez, Roman V. Palomares, Eulalio Chavez, Matilde Trevino y otros. Estos siervos de Dios fueron las columnas en el Templo en aquellos dias, y el nucleo mas distinguido y formal que haya tenido nuestra iglesia en las filas de su ministerio en los primeros anos de la Iglesia Metodista Mexicana (palabras del Hno. Pablo G. Verduzco)

En 1886 se establecio en Monterrey, Nuevo Leon un Colegio para educar a los jovenes aspirantes al ministerio cristiano, bajo el nombre de Instituto Fronterizo y bajo la direccion del Prof. P. C. Bryce. En este Instituto se prepararon Pedro Grado, autor de varios de nuestros himnos, Andres Osuna, muy conocido y muy estimado por su consagracion y relevantes dotes intelectuales, Julian Castro, Ruperto Revino y otros. Los primeros tres fueron excelentes oradores sagrados, y los tres llegaron a ser ordenados presbiteros. Este mismo ano de 1886 se reunio la Conferencia anual en este Instituto bajo la presidencia del Sr. Obispo Jose S. Key. Estuvieron presentes los Revdos. Alejandro Sutherland, Alejandro de Leon, Santiago Tafolla, J. P. Rodriguez, Eulalio Chavez, Basilio Soto, Santos Romo y otros. Estos obreros fueron los veteranos de la antigua guardia, los fundadores de nuestra Iglesia.

Borquejo Históricos de Metodismo Mexicano
Pablo García Verduzco.

Historical review of the Methodist Church.

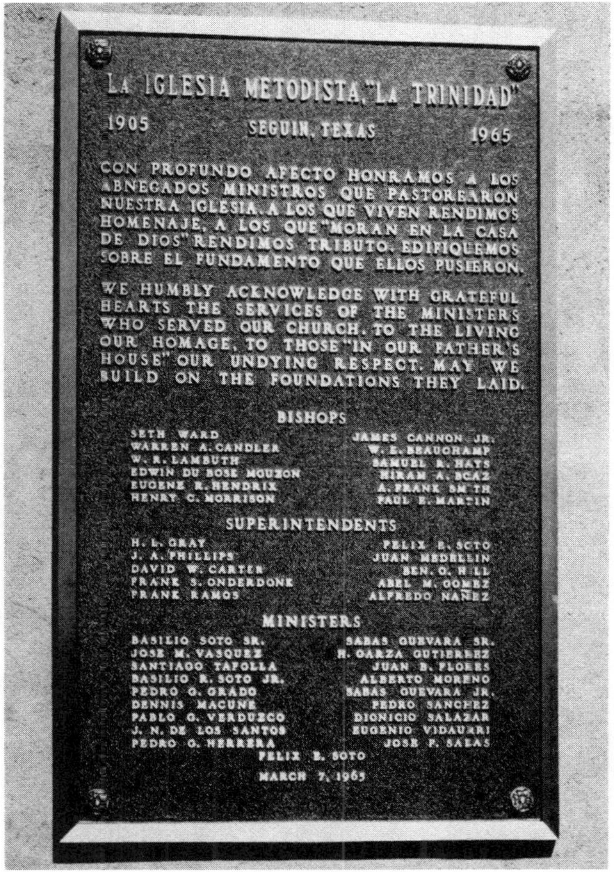

Memorial Plaque commemorating 60th anniversary of La Trinidad Methodist Church, 1965, listing Santiago as one of the early pastors.

Camp meeting, 1925, with Rev. M.F. Tafolla at right.

A Tafolla family gathering in the 1930s. Two of Santiago's sons are present, Gabriel and Mariano, and many grandchildren and great-grandchildren.

The Family of Rev. Mariano F. Tafolla and Eloisa Sánchez Tafolla c. 1924. M.F. Tafolla, Santiago's son, was Fidel's father, and the grandfather of Laura Tafolla and Carmen Tafolla. Seated front (left to right): Mariano, Jr. (father of Carmen Tafolla), M. F. Tafolla, Sr., Margie, Eloisa, Robert B., Evangelina. Standing (left to right): Sarah, Phoebe, Eloisa, Fidel L. (father of Laura Tafolla), Esther, Sofía, Anita.

1018 El Monte Blvd.
San Antonio, Texas

December 7, 1970

Mr. Alfredo Nanez, Director
Perkins School of Theology
Southern Methodist University
Dallas, Texas

Dear Brother Nanez:

I have visited twice in the home of Mr. and Mrs. W. W. Jackson, the last time the early part of this week. I submitted a copy of my grand-father's auto-biography for their inspection and they were very complimentary about it, saying that the story should be preserved for posterity. Mr. Jackson told me that you were now in a better position to talk to the higher authorities of the Methodist church about finding a way to publish it.

Since I wrote you last, I have translated it to English, have made a trip to Santa Fe and two trips to Bandera to authenticate the story. I also visited the church in Seguin and verified the date of his death at the Court House. He died October 14, 1911 just three years after he wrote his story at Pearsall, Texas. His first wife was a cousin to Polly Rodriguez. You are correct about 1875 being a turning point in his life. He and Polly were comrades-in-arms, very close friends, and lived in the Privilege Creek Community. James Tafolla, as you say, became a member of the Missionary Conference in Bandera and was recommended to the Annual Conference which met at Seguin on October 25th, 1876. It is interesting to note that he died in the same city and the same month thirty-five years later.

According to church records, he assumed his pastoral duties in March, 1877. You have been very helpful in sending me his "Oja de Servicio". I have filled in some of the blanks and shall keep trying to fill in the others. In the meantime, if you run across any other information, please fill it in. I enclose a zerox copy of the sheet which still has blanks.

I have finished translating the story into English so I have now the English and the Spanish version. The material covers about 75 pages of doubled space typing (one side of paper only).

It is almost impossible to evaluate the story to find out if it could be used as a part of church history without reading it. If you have the time to read it, I shall be glad to zerox a copy to send to you. You could send it back to me when you get through.

I want to express my gratitude to you for your encouragement and interest in the matter. Please give my regards to your wife and may God bless you in the work you have now undertaken.

Yours in Christ,

F. L. Tafolla

FLT/hf

enc.

Fidel's letter to Alfredo Nanez, December 7, 1970.

1018 El Monte Blvd.
San Antonio, Texas

December 7, 1970

The Readers Digest Ass'n., Inc.
Plesantville, New York, 10570

Dear Sirs:

After serving 50 years in the public schools of San Antonio, Texas, I retired as principal of one of the high schools in May, 1968.

My grandfather, James Tafolla Sr., wrote his memoirs in 1908 just three years before he died in 1911. The manuscript was circulated among the members of his family for a year or so, then one of his sons put it in a steel safe where it remained in its original form approximately sixty years. It was never copied or typed. All his sons and most of his grandchildren have passed away. I am the oldest of the grandsons living and the members or descendents of the family have urged me to edit the story for publication.

I have been a reader of the Readers Digest for many years and I think the story would be good material for your Book Section.

Briefly it is the story of a boy who ran away from his home in Santa Fe, New Mexico, because of the way he was mistreated by an older brother. He was 12 years old then and the story begins in 1848. He joined a wagon train that had carried freight to Santa Fe and was returning by way of the Santa Fe Trail to Independence, Missouri. From Independence, Mo., his benefactor and sponsor took him by stagecoach all the way to New York and back to Washington, D. C. and other large cities along the Eastern seaboard. He describes in detail, the hardships and adventures on the road, the prairie dog towns, the buffalo and the precautions taken when the train camped for the night. He tells his adventures, his struggles to learn English and to make a living. At the age of 17 he joined the Second U. S. Cavalry hoping to get back to his native state. His regiment, however was ordered to Texas to fight Indians, and he tells of the hardships on the expeditions to fight Indians and how he was wounded by an arrow in one of the Indian skirmishes. He mustered out of the Cavalry in 1860. Soon after this he joined the Confederacy. He relates his experiences in this service and how close to the end of the conflict he and four of his friends deserted, cut across undeveloped land avoiding the known roads and finally succeeded in crossing the Rio Grande into Mexico. He tells how they were received and what he did for a living while there. After the Civil War ended he came back to Texas and lived the life of a pioneer in Bandera, County, Texas.

The whole story has a great deal of human interest, pathos, and local color and makes interesting reading. It has a special appeal for the student of early life in the United States and Texas in particular.

Please let me know if you want me to submit a copy of the manuscript (typed) for your consideration and if the manuscript is returned to me. Also please let me know on what terms you accept a story or award a contract.

Hoping to hear from you at an early date, I remain,

Sincerely yours,

F. L. Tafolla (Editor)

FLT/hf

P.S.
The original manuscript was written in Spanish, with pencil and on very cheap tablet paper. It is in my possession now, but is very fragile. I had it typed and made the necessary corrections in Spanish. Then I translated it into the English language, so I have the Spanish and the English versions of the story. It covers 75 pages of double spaced typed material (one side of paper).

I have made a trip to Santa Fe, New Mexico to authenticate material, get pictures, footnotes, etc. It is almost ready to be printed.

Fidel's letter to Reader's Digest, December 7, 1970.

Franciscan Fathers
Guadalupe Church and Indian Missions
Pena Blanca, New Mexico 87041
Phone Pena Blanca 465-2226

May 12, 1970

Dear Mr. Tafolla:

Your letter and enclosures, dated June 1, were forwarded to me but arrived today, well after previous communications -- and the letter I wrote you recently.

This letter confirms what I had suspected in the other communications received, that you wanted connections with the Tafoya Altamirano family of New Mexico.

As I wrote to you, I have neither the time nor opportunity to pursue genealogical tracings for anyone. Undoubtedly, your grandfather belonged to the Tafoyas of Santa Fe.

Microfilms of church registers prior to 1850 are at the State Hall of Records, Santa Fe, New Mexico. Someone there might be able to find what you want, aorund 1837, if you can give him the names of the parents of Santiago Tafoya.

Sincerely,

Fr. Angelico Chavez, OFM

State Archives
404 Montezuma

Response to Fidel from Fray A. Chavez, May 12, 1970.

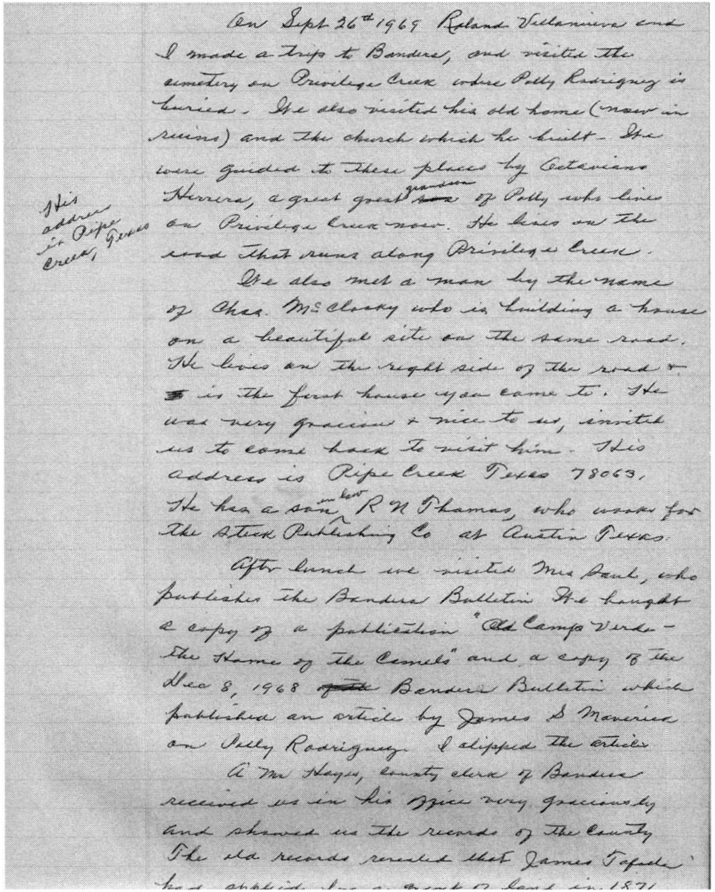

On Sept 26th 1969 Roland Villanueva and I made a trip to Bandera, and visited the cemetery on Privilege Creek where Polly Rodriguez is buried. We also visited his old home (now in ruins) and the church which he built. We were guided to these places by Octaviano Herrera, a great great grandson of Polly who lives on Privilege Creek now. He lives on the road that runs along Privilege Creek.

His address is Pipe Creek, Texas

We also met a man by the name of Chas. McClosky who is building a house on a beautiful site on the same road. He lives on the right side of the road & is the first house you come to. He was very gracious & nice to us, invited us to come back to visit him. His address is Pipe Creek Texas 78063. He has a son in law R N Thomas, who works for The Steck Publishing Co at Austin Texas.

After lunch we visited Mrs [illegible], who publishes the Bandera Bulletin. We bought a copy of a publication "Old Camp Verde – the Home of the Camels" and a copy of the Dec 8, 1968 Bandera Bulletin which published an article by James S Maverick on Polly Rodriguez. I clipped the article.

A Mr Hayes, county clerk of Bandera received us in his office very graciously and showed us the records of the County. The old records revealed that James Tafolla had applied for a grant of land in 1871

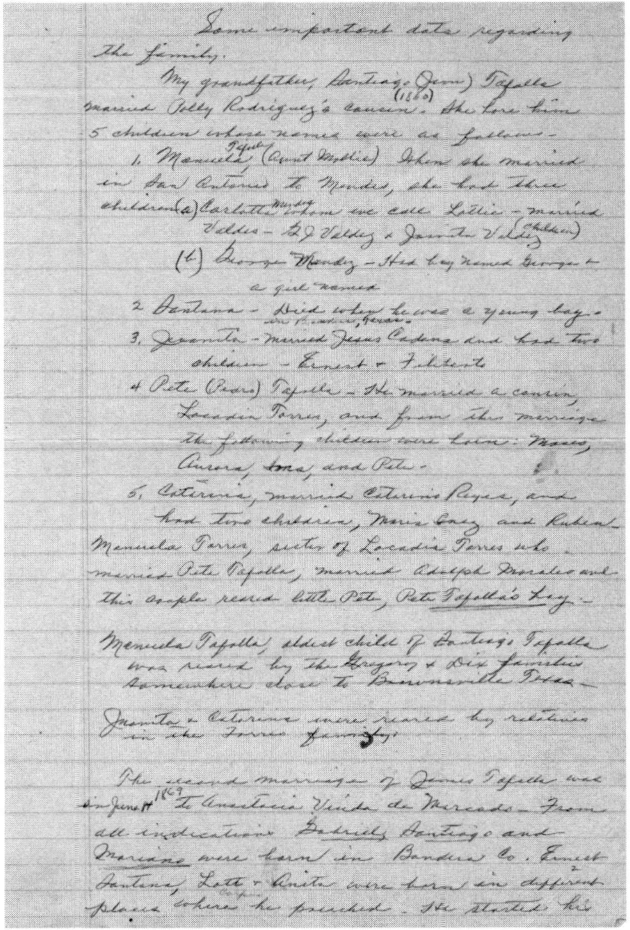

Some important data regarding the family.

My grandfather, Santiago (James) Tafolla married Polly Rodriguez's cousin (1860). She bore him 5 children whose names were as follows –

1. Manuela Tafolla (Aunt Mollie) When she married in San Antonio to Mendez, she had three children (a) Carlotta Mendez whom we call Lottie – married Valdez – E J Valdez & Juanita Valdez (children)
(b) George Mendez – Had boy named George & a girl named

2 Santana – Died when he was a young boy in Bandera, Texas.

3. Juanita – married Jesus Cadena and had two children – Ernest & Filiberto

4 Pete (Pedro) Tafolla – He married a cousin, Locadia Torres, and from this marriage the following children were born: Moses, Aurora, Ima, and Pete.

5. Caterina, married Catarino Reyes, and had two children, Maria Cruz and Ruben.

Manuela Torres, sister of Locadia Torres who married Pete Tafolla, married Adolph Morales and this couple reared little Pete, Pete Tafolla's boy.

Manuela Tafolla, eldest child of Santiago Tafolla was reared by the Gregory & Dix families somewhere close to Brownsville Texas.

Juanita & Caterina were reared by relatives in the Torres family.

The second marriage of James Tafolla was in June 11th 1869 to Anastacia Viuda de Mercado. From all indications Gabriel, Santiago and Mariano were born in Bandera Co. Ernest, Santana 2, Lott & Anita were born in different places where he preached. He started his

Fidel's handwritten notes while researching Santiago's life.

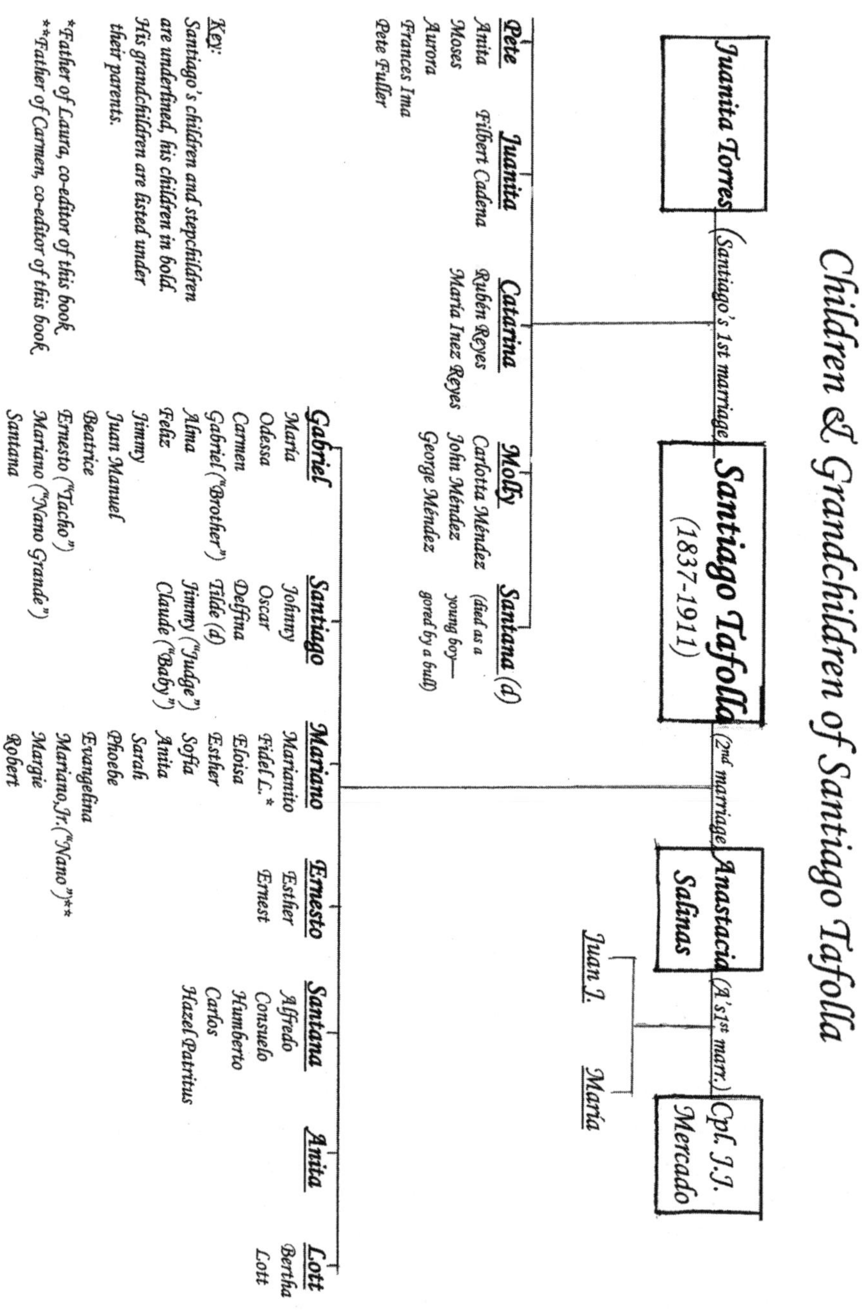

Children and grandchildren of Santiago Tafolla.

ACKNOWLEDGEMENTS

Over the last 100 years, there have been many people who have contributed to the realization of this book, from family members who treasured this story to librarians and community activists, like Martha Cotera, brilliant author of the Rockefeller Grant brainchild to preserve Mexican-American Archives at UT Austin, who first insisted in the early 1980s that a copy of this story be archived at the Nettie Lee Benson Latin American Collection. Many have helped and lent a hand at each point along the way, and to all of you we are grateful. In 1973, Oscar Tafolla, grandson of Santiago, and son of Santiago Jr., entrusted Carmen, his cousin's child, with the keeping of documents and stories about Santiago and asked that she make sure they not be lost to history. And to Consuelo, Julia, Bronco Pete, Mariano F., Gabe, Sarah, Jimmy, Oscar, Mariano, Jr., Bobby, Tacho, Joe, Gil, Ana, Mary, Esther, Rosie, Alma, and all of the Tafollas who have added a piece of information, a newspaper article, a story, an enthusiastic ear, or faith in the power of this story, we are deeply indebted.

But in 2007, an enthusiastic scholar hard at work on her doctoral dissertation approached us about using our great-grandfather's unpublished manuscript in her work. Apparently, while doing her research, one of the thirty or so typewritten copies of Fidel's translation had fallen across her path and she wanted to know more about the Spanish original. Norma Mouton's commitment to the importance of Santiago's work sparked our renewed energy on a project that had long been on our to-do list, overshadowed by the demands of everyday life. She encouraged us to meet with Dr. Nicolás Kanellos, publisher of Arte Público Press, a significant source for the publication of Hispanic literary heritage documents. We were impressed with the priority and care they had given documents of historic importance and literary interest that had previously not been accessible to professional research, but stored instead in someone's basement or in an old trunk. Thank you, Norma, for paving the way for this publication's appropriate placement. It is because of you and your graciousness and integrity, that we re-prioritized our professional schedules, and committed ourselves to the urgent preparation of this work. Your ability to use this document in your dissertation as a formative portion of the research on the literature of conversion (Congratulations, Dr. Mouton!) also revealed the timeliness of making this document available to historians, cultural researchers, scholars and courageous rebels everywhere.

The selection of Arte Público as the publisher of this work turned out to be not only appropriate but most gracious. Gabriela Baeza Ventura, Ph.D., Senior Editor for Arte Público and Associate Professor of Hispanic Literature for the Department of Hispanic Studies, provided leadership as a dedicated and talented scholar along the road of this wondrous, eye-opening, nerve-wracking journey. Dr. Kanellos has lent support, enthusiasm, and perspective to our research endeavors. So many of the AP staff were our unseen heroes and even though your name is not mentioned here, we appreciate your efforts on our behalf. We must pay homage also to Carolina Villarroel, Ph.D., Brown Foundation Director of Research for the University of Houston's Recovering the U.S. Hispanic Literary Heritage Project for her untiring service, and also for her guided tour of the second floor archives. Every time we had a question or needed help copying documents or pictures, Carolina and Gabi were there for support and advice. Many thanks also to Marina Tristán, Assistant Director, Arte Público Press, for the beautiful and concise catalog description of this work, and for enabling us to arrive at a title which would capture the essence of this great man's life.

Along the way, we have received counsel, help and insight from friends, strangers, and loved ones alike. Thank you to Wayne Vick, our local Civil War enthusiast who added facts, dates and timelines that corresponded to the story. This greatly enhanced our understanding of the locations and events that Santiago referred to in the document. Wayne remarked that when he first received one of the typewritten copies of Fidel's translation, "time stood still" and he dropped everything to read it. This comment has continued to fuel our desire to present this work to the public. Thank you to Dr. Antonia Castañeda for her literary and historical input and wise advice. A huge debt of gratitude to the patience and tedious professional bibliographic research by a promising young writer, Alexandra Araiza. Thank you to Tom Dugger who rescued lost data free-of-charge from an unruly hard-drive which crashed during a sudden storm. To cousins Henry and Kathy Dominguez who did much of the legwork during that tense time of uncertainty, and to our friends who listened to all the details and gave their feedback and support, we appreciate having you in our lives.

From Laura: A special "shout-out" to Diane and John Hardin, and Amy Orosco for your loyal friendship. And to my family who put up with my cranky, stressed out rants, and all night cookie-and-milk computer sessions. Thank you Alicia Machado, Bobby Ramirez, Abel and Janie Ramirez, Joanne Treviño and Robert Alexander for making me laugh when I wanted to cry.

From Carmen: To my soul-partner, Ernesto M. Bernal, who listened patiently with an editor's ear in English, Spanish, and the intergalactic language of scholarly documentation, and to my loving children Israel, Ariana, and Mari, thanks for your patience when I said, "Just ten minutes more—I'm in the middle of the Civil War!" and then proceeded to go another ten minutes and another and another. Israel, your scanning and technical assistance was a godsend, and Alexandra Araiza, your research assistance and brilliance of detail helped make this book possible. And to my mother, father, and Aunt Sarah, who took the time to tell me many, many of the family stories.

We wish to acknowledge with highest regard and utmost respect our great-grandfather, Santiago, who at the age of 71 and in the middle of his very busy work schedule, had the presence of mind to sit down and write, in remarkable detail, the fascinating story of his life. After sitting with the handwritten tablets for a while, it was evident that the words had flowed freely out of him as if he had been urgently inspired. During our work on this book, we grew to love him as surely as if we had grown up listening to those stories at his knee, marveling at his courage, intelligence, and independence. We are proud of the man he chose to be and hope that he would be pleased with how his story has been presented. His strength and faith have become ours, *gracias a Dios*.

Most importantly, were it not for the time, talents and dedication of Fidel León Tafolla, these faded tablets may have remained in a dusty white shirt box or been lost to history, never having seen the light of day.

Along with Santiago, you have been our inspiration through this entire process. Your unseen hands have led us to folders and files and articles that have helped us chronicle the events, and have given solid documentation of our great-grandfather's story. Thank you Fidel for the passion you had so many years ago of sharing your grandfather's fascinating escapades with all of us. We are proud to say that your dream has finally become a reality.

Laura Tafolla, Editor
Carmen Tafolla, Editor

ABOUT THE TRANSLATOR

Santiago's grandson Fidel (son of Mariano, Sr.) was the original translator of this document. Born in 1898, he was 13 years old when Santiago died. Raised in a family of educators and storytellers, he understood the idiosyncrasies and innuendos of Santiago's language, in both English and Spanish, was familiar with his usages and references, and had heard the oral tales many times.

Fidel himself was a man of many accomplishments. The oldest survivor of a family of twelve siblings, he struggled to gain a college education in a time period when few Mexican-Americans were even able to enter high school. He worked as a teacher, while still trying to gain his degree, and earned, not only a Bachelor of Arts, but also a Masters Degree from Texas State. Following a long-standing family tradition of literate bilingualism, Fidel made a mark on his chosen profession, not only as a beloved Spanish teacher at Main Avenue (now known as Fox Tech) and Thomas Jefferson High Schools in San Antonio, but also as an agent of change. He was the first Hispanic president of the San Antonio Teachers' Council back in 1934. He was also instrumental in starting the San Antonio Retired Teachers' Association and later the San Antonio Teachers' Credit Union (now operating as FirstMark Credit Union). He became Dean of Boys at Lanier High School, then Vice-Principal and finally Principal.

Serving as an elder in the Presbyterian Church, a member of the Master Masons, and in later years, a Shriner, Fidel was known as a deeply spiritual man, but also as a man with a great sense of humor. He was a faithful mentor to many, and the devoted father of five children: Dorothy Jane, Elizabeth Ann, Fidel Jr., Donald Edward and Laura Jean.

After teaching for fifty years, most of them at Lanier High School in the predominantly Mexican-American West Side of San Antonio, Fidel retired in 1968, and the San Antonio Independent School District chose to honor him by naming a new campus across the street from Lanier H.S. after him. Attending the dedication of the new Fidel Tafolla Jr. High School with his family was one of his proudest moments.

Upon retirement, he set to work on an idea that had been developing within him for years—to achieve the publication of his grandfather Santiago's autobiography. He contacted scholars and archivists, and began to translate and prepare the manuscript for publication. However, in the late 1960s there was still insufficient interest on the part of mainstream publishers in the story of a Mexican American who did not even fit the stereotype of what "Mexicans" were "known" to be. Fidel was still attempting to find a publisher for the story in January of 1971 when he died of a sudden heart attack, not unlike the one which had taken Santiago almost sixty years before.

Fidel Tafolla

ABOUT THE EDITORS

Carmen Tafolla is the great-granddaughter of Santiago Tafolla. She first began writing down Tafolla family history when she was 13 years old, after hearing the stories her Aunt Sarah told her about Santiago. Despite the number of years that lapsed between that point and the publication of this book, she has not been totally idle. Her work has appeared in more than 200 anthologies, and she has published ten children's books, five books of poetry, seven television screenplays, a book of short stories, and many articles and essays. Carmen holds a Ph.D. from the University of Texas, and carries on the family tradition of literate bilingual biculturalism. She has performed her one-woman show of Mexican-American voices to standing ovations in England, Spain, Mexico, Norway, Germany, Canada, New Zealand, and throughout the United States.

In 1999, Carmen was awarded the Art of Peace Award for work that contributes to peace, justice, and human understanding, and has been recognized by the National Association for Chicano Studies for work which "gives voice to the peoples and cultures of this land." She is the recipient of numerous awards, including the Charlotte Zolotow Award for best children's picture book writing of 2010, The 2009 Tomas Rivera Book Award for *The Holy Tortilla and a Pot of Beans*, the 2010 Tomás Rivera Book Award, the 2010 Américas Award, and two International Latino Book Awards for *What Can You DO With a Paleta?*, and ALA Notable and Junior Library Guild Selection for *What Can You DO With a Rebozo?* Her writings are archived at the Benson Latin American Collection of the University of Texas Libraries, and she is currently at work on the biography of noted labor organizer Emma Tenayuca. Like Santiago, Carmen is still crossing borders, and trying to break down the barriers between groups of people; unlike him, she did not inherit the internal geographic compass of a scout and is often seen headed the wrong way on Texas highways and in airport terminals. You can find more information on her at www.carmentafolla.com.

Laura Jean Tafolla is the great-granddaughter of Santiago Tafolla. She is the daughter of Olivia Treviño Tafolla and Fidel León Tafolla, the original translator of this manuscript.

Since both of her parents were educators—her mother was an elementary teacher, her father a high school principal—it was a natural choice for Laura to dream of becoming a teacher. She majored in Psychology and Education with a minor in Mathematics and received a Bachelors Degree from Trinity University in 1976. At Trinity, she was inducted into Kappa Delta Pi, the education honor society. She also received a Masters in Educational Psychology and Counseling from the University of Texas at San Antonio in 1980.

Laura began teaching in 1976 and has taught in the San Antonio Independent School District for more than 30 years. She spent most of her career at Highlands High School as a teacher of mathematics, but has also taught at Lowell Middle School and Tafolla Middle School, her father's namesake, and has worked as an independent counselor with several social agencies. Laura has retired as a full-time educator, but works part-time in the Gear Up Program as a math tutor at Lanier High School. She had always wanted to teach at the same school where her father taught, and was thrilled when the opportunity arose.

Never one to be limited by titles, and true to her artistic genes, Laura is also an accomplished singer, and in her spare time enjoys painting and writing. Upon discovery of Santiago's 100-year-old, handwritten manuscript (in an old shirt box in the closet) her father's dream became her own. She is excited to share this captivating work with the world.